Oxford Series on Materials Modelling

Materials modelling is one of the fastest growing areas in the science and engineering of materials, both in academe and in industry. It is a very wide field covering materials phenomena and processes that span ten orders of magnitude in length and more than twenty in time. A broad range of models and computational techniques has been developed to model separately atomistic, microstructural, and continuum processes. A new field of multiscale modeling has also emerged in which two or more length scales are modeled sequentially or concurrently. The aim of this series is to provide a pedagogical set of texts spanning the atomistic and microstructural scales of materials modeling, written by acknowledged experts. Each book will assume at most a rudimentary knowledge of the field it covers and it will bring the reader to the frontiers of current research. It is hoped that the series will be useful for teaching materials modeling at the postgraduate level.

APS, London
RER, Livermore, California

1. M. W. Finnis: *Interatomic Forces in Condensed Matter*
2. K. Bhattacharya: *Microstructure of Martensite—Why It Forms and How It Gives Rise to the Shape-Memory Effects*
3. V. V. Bulatov, W. Cai: *Computer Simulations of Dislocations*
4. A. S. Argon: *Strengthening Mechanisms in Crystal Plasticity*
5. L. P. Kubin: *Dislocations, Mesoscale Simulations and Plastic Flow*
6. A. P. Sutton: *Physics of Elasticity and Crystal Defects*
7. D. Steigmann: *A Course on Plasticity Theory*

Forthcoming:
D. N. Theodorou, V. Mavrantzas: *Multiscale Modelling of Polymers*

OXFORD SERIES ON MATERIALS MODELLING

A Course on Plasticity Theory

David J. Steigmann

Department of Mechanical Engineering, University of California, Berkeley

OXFORD
UNIVERSITY PRESS

OXFORD
UNIVERSITY PRESS

Great Clarendon Street, Oxford, OX2 6DP,
United Kingdom

Oxford University Press is a department of the University of Oxford.
It furthers the University's objective of excellence in research, scholarship,
and education by publishing worldwide. Oxford is a registered trade mark of
Oxford University Press in the UK and in certain other countries

Impression: 1

Published in the United States of America by Oxford University Press
198 Madison Avenue, New York, NY 10016, United States of America

British Library Cataloguing in Publication Data
Data available

Library of Congress Control Number: 2022942494

ISBN 978–0–19–288315–5
DOI: 10.1093/oso/9780192883155.001.0001

Printed and bound by
CPI Group (UK) Ltd, Croydon, CR0 4YY

To my family

Preface

The theory of plasticity has a long and interesting history dating back about two and a half centuries. Activity in the field expanded rapidly over the course of the past century in particular, giving rise to a rapid pace of advancement. During much of the latter phase of its modern development, the field was beset by ambiguity and controversy concerning some of its conceptual foundations. Unsurprisingly, this led to the emergence of different, often incompatible, schools of thought on the subject. A comprehensive survey of the state of plasticity theory during this period may be found in the review article by Naghdi. Meanwhile, great strides were being made by applied mathematicians in laying the foundations of modern continuum mechanics. Their emphasis on permanence and rigor meant that the unsettled subject of plasticity theory was largely avoided, however, with the result that this lacuna in the panoply of continuum theories began to be filled quite recently, around the turn of the millennium, after the antagonism of the older schools had begun to fade.

Practically everything known about plasticity through the middle of the past century is documented in the superb treatises by Prager and Hodge, Nadai, Hill, and Kachanov, which should be carefully read by any serious student of our subject. At around the same time, new developments were taking place in the application of differential geometry to the continuum theory of defects associated with plasticity. This has become a large and active discipline in its own right, and a substantial part of this book is devoted to it. The works of Bloom and Wang, and the volume edited by Kröner, are recommended to those interested in learning about its foundations, while those by Clayton, Epstein and Elżanowski, Epstein, and Steinmann cover many of the more recent developments. The modern engineering theory, as distinct from the geometrical theory, is ably summarized in the books by Lubliner, Besseling and van der Giessen, and Bigoni. The books by Han and Reddy and by Gurtin et al. are recommended for mathematical developments and some of the more recent thinking on the subject.

While writing this book I have been guided by the belief that one can always learn something from any thoughtful person. Accordingly the contents reflect my understanding of the work of researchers and scholars spanning a large and diverse range of views on the subject of plasticity. In the course of surveying the modern literature, I have been struck by the continuing isolation of the various schools from one another, with scant evidence of cross-fertilization. Particularly glaring, from my perspective, is the lack of acknowledgment of the efforts of Noll in laying the foundations of the modern theory. This has been rectified to a great degree by Epstein and Elżanowski, and I follow their lead in giving primacy to Noll's perspective. In fairness, Noll is not an easy read, and much study is needed to grasp the full import of his work.

The book is certainly not self-contained. Readers are presumed to have had prior exposure to a good introductory course on basic continuum mechanics at the level of the excellent books by Chadwick and Gurtin, for example. Aspects of this basic background are summarized as needed, but not developed in any detail. The emphasis here is on conceptual issues concerning the foundations of plasticity theory that have proved challenging, to me at least. These have led me to the view that the time has come to seek a measure of consolidation and unification in the field. I do not ignore the classical theory, but rather develop it from the perspective of the modern theory. For example, the classical theory of perfectly plastic solids was presented historically in a way that led to its natural interpretation, from the vantage point of modern continuum mechanics, as a theory of non-Newtonian fluids rather than as a model of the behavior of certain solids. The resolution of this dilemma is a prime example of the clarity that can be achieved once a secure logical foundation for the general theory has been established.

Some explicit solutions to the equations of plasticity theory are covered in this book, but not nearly to the extent found in the older books. The reason for this omission is, firstly, that the small collection of explicit solutions that are known is ably covered elsewhere, so that duplication is hardly justified, and secondly, that due to the advent of modern computing, they are not nearly as relevant as they once were. I devote the remainder of the book to the theoretical foundations of the subject, in accordance with my own predilections, rather than to matters having to do with computation. The reason for this emphasis is my belief that students are typically not as well versed in the conceptual foundations as they should be if they are to realize the full potential of computational mechanics. A number of exercises of varying degrees of difficulty appear throughout. These serve to reinforce understanding and to encourage the reader to fill in any gaps in the development. Comprehensive solutions to selected exercises are included at the end of the book.

Those who might have read my previous book, *Finite Elasticity Theory*, will find the style and presentation of this one to be quite familiar. The present book is perhaps a bit more demanding, however, insofar as various concepts from non-Euclidean differential geometry are covered in detail. I gratefully acknowledge the small group of dedicated graduate students at the University of California, Berkeley, whose interest and persistence provided the impetus for the development of a graduate course on which the book is based. I am especially grateful to one of them, Milad Shirani, for his critical reading of the manuscript and for preparing the figures.

David Steigmann
Berkeley, 2021

References

Besseling, J. F., and van der Giessen, E. (1994). *Mathematical Modelling of Inelastic Deformation.* Chapman and Hall, London.

Bigoni, D. (2012). *Nonlinear Solid Mechanics: Bifurcation Theory and Material Instability.* Cambridge University Press, Cambridge, UK.

Bloom, F. (1979). *Modern Differential Geometric Techniques in the Theory of Continuous Distributions of Dislocations*. Lecture Notes in Mathematics, Vol. 733. Springer, Berlin.

Chadwick, P. (1976). *Continuum Mechanics: Concise Theory and Problems*. Dover, New York.

Clayton, J. D. (2011). *Nonlinear Mechanics of Crystals*. Springer, Dordrecht.

Epstein, M. (2010). *The Geometrical Language of Continuum Mechanics*. Cambridge University Press, Cambridge, UK.

Epstein, M., and Elżanowski, M. (2007). *Material Inhomogeneities and Their Evolution*. Springer, Berlin.

Gurtin, M. E. (1981). *An Introduction to Continuum Mechanics*. Academic Press, Orlando.

Gurtin, M. E., Fried, E., and Anand, L. (2010). *The Mechanics and Thermodynamics of Continua*. Cambridge University Press, Cambridge, UK.

Han,W., and Reddy, B. D. (2013). *Plasticity: Mathematical Theory and Numerical Analysis*. Springer, N.Y.

Hill, R. (1950). *The Mathematical Theory of Plasticity*. Clarendon Press, Oxford.

Kachanov, L. M. (1974). *Fundamentals of the Theory of Plasticity*. MIR Publishers, Moscow.

Kröner, E. (Ed) (1968). *Proc. IUTAM Symposium on Mechanics of Generalized Continua*. Springer, N.Y.

Lubliner, J. (2008). *Plasticity Theory*. Dover, N.Y.

Nadai, A. (1950). *Theory of Flow and Fracture of Solids*. McGraw-Hill, N.Y.

Naghdi, P. M. (1990). A critical review of the state of finite plasticity. *J. Appl. Math. Phys.* (ZAMP) 41, 315–394.

Noll, W. (1967). Materially uniform simple bodies with inhomogeneities. *Arch. Ration. Mech. Anal.* 27, 1–32.

Prager, W., and Hodge, P. G. (1951). *Theory of Perfectly Plastic Solids*. John Wiley & Sons, N.Y.

Steigmann, D. J. (2017). *Finite Elasticity Theory*. Oxford University Press, Oxford.

Steinmann, P. (2015). *Geometrical Foundations of Continuum Mechanics: An Application to First and Second-Order Elasticity and Elasto-Plasticity*. Lecture Notes in Applied Mathematics and Mechanics, Vol. 2. Springer, Berlin.

Wang, C. -C. (1979). *Mathematical Principles of Mechanics and Electromagnetism. Part A: Analytical and Continuum Mechanics*. Plenum Press, N.Y.

Contents

1

Preliminaries

We begin with a fairly descriptive discussion of the main observations about plastic behavior and the basic mechanisms responsible for it. This is followed by a brief resumé of the standard continuum theory that underpins our subsequent development of a theoretical framework for the description of elastic-plastic response.

1.1 Phenomenology

Much of the basic phenomenology of plasticity can be understood in terms of a simple tension-compression test on a uniform metallic bar. Suppose the bar has length l_0 in its unloaded state, and let $T = F/A$ be the uniaxial Cauchy stress in the direction of the bar axis, where F is the axial force and A is the cross-sectional area of the deformed bar. The *stretch* of the bar, presumed to be strained homogeneously, is $\lambda = l/l_0$, where l is the bar's length when deformed. If the stress is not too large, the response of the bar is typically well described by the linear relation

$$T = E \ln \lambda \tag{1.1}$$

between the stress and the logarithmic strain $\ln \lambda$, in which the proportionality constant E is Young's modulus—a property of the material of which the bar is made. This relation presumes the state of the bar, as determined by the stretch and the Cauchy stress, to be uniform. The bar is then in equilibrium insofar as the effects of body forces (e.g., the weight of the bar) can be neglected.

The range of stresses for which this relation holds is limited. It fails when the stress reaches certain limits, called the *yield stresses* in uniaxial tension or compression. Often these limits coincide in magnitude, so that (1.1) is valid provided that

$$|T| < T_Y, \tag{1.2}$$

where T_Y, another property of the material, is the *initial* yield stress, the qualifier reflecting the fact that the yield stress usually evolves with the state of the material under continued deformation, its current value typically exceeding the initial value. This phenomenon, called *strain hardening*, is depicted schematically in Figure 1.1. If the bar is

A Course on Plasticity Theory. David J. Steigmann, Oxford University Press. © David J. Steigmann (2022).
DOI: 10.1093/oso/9780192883155.003.0001

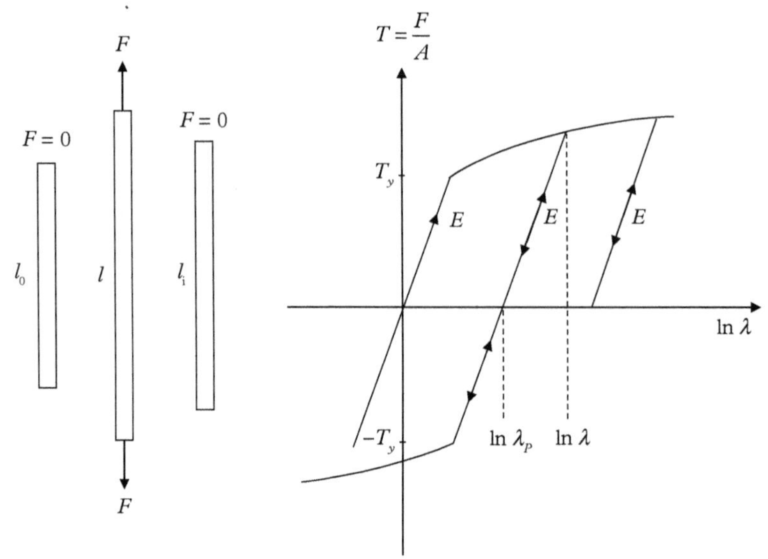

Figure 1.1 *Uniaxial stress–strain response of a bar.*

unloaded to zero stress from a state in which the value of $|\ln\lambda|$ exceeds that associated with initial yield, then the bar does not return to its initial length l_0, but rather to an intermediate length l_i. We say, rather loosely, that the bar has been permanently deformed. The *plastic stretch* associated with l_i is $\lambda_p = l_i/l_0$. Further, the slope of the unloading curve is approximately constant and equal to that of the loading curve, namely E.

The stretch λ of the bar just prior to unloading is thus given by

$$\lambda = \lambda_e\lambda_p, \tag{1.3}$$

where $\lambda_e = l/l_i$ is the *elastic stretch*, so named because, according to the graph,

$$T = E(\ln\lambda - \ln\lambda_p) = E\ln\lambda_e. \tag{1.4}$$

Thus, the elastic stretch bears the same relation to the stress as that associated with the initial elastic response of the bar.

The phenomenology just described leads immediately to the important observation that the elastic properties of the material, as reflected in the uniaxial case by Young's modulus, are roughly insensitive to plastic deformation. This observation carries over to other elastic properties of crystalline materials, as documented in the extensive experimental work of G. I. Taylor and associates and summarized in the introductory chapter of Hill's classic treatise. In particular, the basic lattice structure of a metallic crystal, the seat of its elastic properties, remains largely undisturbed by the relative plastic slip of crystallographic planes. This observation will be incorporated as a cornerstone of the theory to be developed.

Beyond this it is invariably true that $T_Y/E \ll 1$ in metals, implying that $|T|/E \ll 1$ and hence that $|\ln \lambda_e| \ll 1$. Accordingly, $T \simeq E\varepsilon_e$, where $\varepsilon_e = \lambda_e - 1$; that is, the elastic strain ε_e is invariably small in magnitude. A further observation about the uniaxial bar test is that λ_p remains unchanged as long as $|T| < T_Y$, the current value of the yield stress just prior to unloading. Indeed the primary purpose of plasticity theory is to describe how λ_p, or, more accurately, its three-dimensional counterpart, evolves when the yield limit is reached. In connection with this it is necessary to have an effective model of strain hardening, this arguably constituting the main open problem of the phenomenological theory. Indeed this aspect of the subject is a principal focus of much contemporary research. Later in the book, we will endeavor to summarize some of the current thinking in this area.

We have mentioned the role of slip along crystallographic planes as a basic mechanism of plastic deformation, giving rise to an overall shear deformation on the macro-scale. This is essentially a frictional process and thus entails the dissipation of energy. The sliding does not take place all at once, but is instead the product of the progressive movement of *dislocations* through the crystal lattice. Clear illustrations of this phenomenon are given in Figure 102.2 in the book by Gurtin et al. and Chapter 1 of the book by Kovács and Zsoldos. Roughly, the movement of a dislocation is initiated by the breaking of an atomic bond between two atoms occupying adjacent layers as they displace relative to each other in response to an applied shear stress, say. A displaced atom then forms a bond with its new nearest neighbor. This process continues in a sequential manner until all the atoms in a given layer are displaced by one lattice spacing relative to those occupying the adjacent layer. The reason why this process occurs via the passage of a dislocation rather than all at once is that the dislocation mechanism requires substantially less effort. This can be readily understood in terms of the famous carpet analogy: Thus, imagine being tasked with the job of displacing a carpet across a floor, all the while maintaining a substantial amount of contact between the two. This is the analog of the relative plastic slip of adjacent planes of the lattice. One can drag the carpet wholly, of course, but it is much easier to create a narrow wrinkle at one end and simply push it across the remaining part of the carpet. The net effect of this procedure is that the entire carpet has been displaced en masse once it has been traversed by the wrinkle.

A dislocation engenders a local distortion of the lattice in the course of its movement along a crystallographic plane. Recalling that the lattice is the seat of the elastic response of the material, it follows that dislocations indirectly induce a local stress field in their vicinity. Thus, to the extent that dislocations are present in an unloaded crystal, they generate a field of *residual stress* in the material. This too is something that a good theory should be able to predict.

Naturally dislocations in real crystals are, like atoms, discrete features, but, like atoms, they are usually so densely distributed in a typical sample as to render meaningful their description in terms of a continuous distribution. We then speak of a *dislocation density* in much the same way as mass density is used to model densely distributed matter. In turn, the notion of a dislocation density has a fascinating connection with certain concepts in non-Euclidean differential geometry, to be explored later.

Most metallic parts used in engineering applications are polycrystalline, consisting of small grains of pure crystal within which the mechanism of dislocation motion is operative. These grains join at grain boundaries, where their interactions contribute to the overall plastic response of the aggregate. Often these grains are more or less randomly oriented, so that at a mesoscopic scale the aggregate responds in the manner of an isotropic continuum. For this reason the classical theory of plasticity is concerned almost exclusively with the response of isotropic materials, whereas theories for crystalline materials are largely confined to the research literature. The volume edited by Teodosiu and the books by Havner and Gurtin et al. are exceptions to this rule and constitute essential reading in the field of crystal plasticity. This is not to say that the theory for isotropic materials is passé. On the contrary, the difficulties encountered in reconciling classical plasticity theory with modern continuum mechanics are readily resolved in the framework of the modern theory. Accordingly we devote substantial space to the isotropic theory in this book.

We confine attention to the purely mechanical theory because this is where the main conceptual challenges lie. Treatments of the thermodynamical theory may be found in the books by Epstein and Elżanowski and by Maugin.

1.2 Elements of continuum mechanics

For the most part, our development is based on the standard framework of continuum mechanics as conceived by Cauchy. Thus, we do not take couple stresses or higher order stresses into account. This is very much in accord with the vast majority of work in plasticity theory. Much of the modern literature also seeks to describe length-scale effects associated with plastic response. This is typically modeled by including gradients of plastic deformation among the variables appearing in constitutive equations, while keeping much of Cauchy's framework intact. We will devote some effort later to a discussion of these developments. For now, however, we shall be content with a brief survey of the basic elements of continuum mechanics that are needed for our work. Detailed discussions of everything said here may be found in the textbooks by Gurtin, Chadwick, and Liu, for example.

Concerning notation, we adopt the standard symbols $\mathbf{A}^t$, $\mathbf{A}^{-1}$, $\mathbf{A}^*$, $Sym\mathbf{A}$, $Skw\mathbf{A}$, $Dev\mathbf{A}$, and J_A. These are, respectively, the transpose, inverse, cofactor, symmetric part, skew part, deviatoric part, and determinant of a second-order tensor $\mathbf{A}$. If $\mathbf{A}$ is invertible, then $\mathbf{A}^* = J_A\mathbf{A}^{-t}$. We also use Sym to identify the linear space of symmetric tensors. The tensor product of 3-vectors is indicated by interposing the symbol $\otimes$, i.e., $\mathbf{a} \otimes \mathbf{b}$, and is defined by $(\mathbf{a} \otimes \mathbf{b})\mathbf{v} = (\mathbf{b} \cdot \mathbf{v})\mathbf{a}$ for any vector $\mathbf{v}$. The Euclidean inner product of tensors $\mathbf{A}$,$\mathbf{B}$ is denoted and defined by $\mathbf{A} \cdot \mathbf{B} = tr(\mathbf{A}\mathbf{B}^t)$, where $tr(\cdot)$ is the trace; the induced norm is $|\mathbf{A}| = \sqrt{\mathbf{A} \cdot \mathbf{A}}$. For a fourth-order tensor $\mathcal{A}$, the notation $\mathcal{A}[\mathbf{B}]$ stands for the second-order tensor resulting from the linear action of $\mathcal{A}$ on $\mathbf{B}$. Its transpose $\mathcal{A}^t$ is defined by $\mathbf{B} \cdot \mathcal{A}^t[\mathbf{A}] = \mathbf{A} \cdot \mathcal{A}[\mathbf{B}]$, and $\mathcal{A}$ is said to possess major symmetry if

$\mathcal{A}^t = \mathcal{A}$. If $\mathbf{A} \cdot \mathcal{A}[\mathbf{B}] = \mathbf{A}^t \cdot \mathcal{A}[\mathbf{B}]$ and $\mathbf{A} \cdot \mathcal{A}[\mathbf{B}] = \mathbf{A} \cdot \mathcal{A}[\mathbf{B}^t]$, then $\mathcal{A}$ is said to possess minor symmetry. The notation $(\cdot)_{\mathbf{A}}$, with a bold subscript, stands for the derivative of a function with respect to tensor $\mathbf{A}$.

Suppose a body B, consisting of a fixed set of material points, occupies a configuration κ_t at time t, a region in a three-dimensional Euclidean space. The restriction to Euclidean space is not sufficiently general to accommodate all conditions. Rather, it reflects a prejudice derived from our terrestrial, non-relativistic, experience, which is nevertheless sufficient to cover most problems that arise at the level of our present technological development. Let $\mathbf{y}$ be the position in κ_t, relative to a specified origin, of a material point $p \in B$. To convey the notion that this position is occupied by p, and only by p, we conceive of an invertible map $\boldsymbol{\chi}$ from B to κ_t such that

$$\mathbf{y} = \boldsymbol{\chi}(p, t). \tag{1.5}$$

Rather than deal with the ethereal body B directly, to facilitate analysis we pick some fixed region of Euclidean space, labeled κ, that stands in one-to-one relation to it. We call this a *reference* configuration. For example, it is usually convenient to choose a region that could, in principle, be occupied by the body, even if it is never actually occupied in the course of its motion. Quite often analysts choose $\kappa = \kappa_{t_0}$, the actual configuration at time t_0, which of course automatically fulfills the occupiability condition. Whatever the choice of the reference configuration, we stipulate that there exists a one-to-one map $\boldsymbol{\kappa}$ from B to κ such that

$$\mathbf{x} = \boldsymbol{\kappa}(p), \tag{1.6}$$

where $\mathbf{x}$ is the position of p in κ relative to some fixed origin. In this way we effectively identify p with the position $\mathbf{x}$ that it occupies in our chosen κ. We then have a one-to-one relation

$$\mathbf{y} = \boldsymbol{\chi}_\kappa(\mathbf{x}, t), \tag{1.7}$$

called the *deformation* of p from κ to κ_t, where

$$\boldsymbol{\chi}_\kappa(\mathbf{x}, t) = \boldsymbol{\chi}(\boldsymbol{\kappa}^{-1}(\mathbf{x}), t), \tag{1.8}$$

in which the subscript is intended to identify our choice of κ. At the risk of being imprecise, we usually suppress it when there is no risk of confusion, and simply write

$$\mathbf{y} = \boldsymbol{\chi}(\mathbf{x}, t), \tag{1.9}$$

with the caveat, of course, that this is *not* the same function as that appearing in (1.5).

We are typically interested in deformations that are continuous and differentiable, meaning that for material points p_1 and p_2 occupying positions $\mathbf{x}_1 = \boldsymbol{\kappa}(p_1)$ and $\mathbf{x}_2 = \boldsymbol{\kappa}(p_2)$, respectively, that are near to each other in κ, there exists a tensor field $\mathbf{F}(\mathbf{x}, t)$, called the *deformation gradient*, such that

$$\chi(\mathbf{x}_2, t) - \chi(\mathbf{x}_1, t) = \mathbf{F}(\mathbf{x}_1, t)(\mathbf{x}_2 - \mathbf{x}_1) + \mathbf{r}(\mathbf{x}_2, \mathbf{x}_1, t), \tag{1.10}$$

where

$$|\mathbf{r}(\mathbf{x}_2,\mathbf{x}_1, t)| = o(|\mathbf{x}_2 - \mathbf{x}_1|), \tag{1.11}$$

in which the Landau symbol $o(\epsilon)$ identifies terms that are smaller than ε for small ϵ; that is, $o(\epsilon)/\epsilon \to 0$ as $\epsilon \to 0$. It then follows that p_1 and p_2 are near to each other in κ_t as well. Equation (1.10) defines the deformation gradient and effectively furnishes the definition of differentiability in this context. In deference to this we often write

$$\mathbf{F} = \nabla\chi, \tag{1.12}$$

to denote the gradient of χ with respect to $\mathbf{x}$. The invertibility of (1.9) implies that $\mathbf{F}$ is an invertible tensor. This is a consequence of the Inverse Function Theorem. See the book by Fleming.

Unfortunately, (1.10) does not afford a useful way to compute the deformation gradient in terms of the function $\chi(\mathbf{x}, t)$. To rectify this, suppose the points p_1 and p_2 are connected by a smooth curve $c \subset \kappa$ with arclength parametrization $\mathbf{x}(s)$, such that $\mathbf{x}_1 = \mathbf{x}(s_1)$ and $\mathbf{x}_2 = \mathbf{x}(s_2)$. Assuming again that these points are near to each other, we then have

$$\mathbf{x}_2 - \mathbf{x}_1 = \mathbf{x}'(s_1)(s_2 - s_1) + \mathbf{o}(s_2 - s_1), \tag{1.13}$$

where $|\mathbf{o}(s_2 - s_1)| = o(s_2 - s_1)$. Combining this with (1.10) gives

$$\tfrac{1}{s_2-s_1}\{\chi(\mathbf{x}(s_2), t) - \chi(\mathbf{x}(s_1), t)\} = \mathbf{F}(\mathbf{x}_1, t)\mathbf{x}'(s_1) + \tfrac{1}{s_2-s_1}\mathbf{o}(s_2 - s_1), \tag{1.14}$$

and passage to the limit $s_2 \to s_1$ yields

$$\chi' = \mathbf{F}\mathbf{x}' \tag{1.15}$$

at $\mathbf{x} = \mathbf{x}_1$, where, with t fixed,

$$\chi'(s) = \tfrac{d}{ds}\chi(\mathbf{x}(s), t). \tag{1.16}$$

Of course this is just the chain rule. In view of (1.9) it is meaningful to write it in the form

$$d\mathbf{y} = \mathbf{F}d\mathbf{x}. \tag{1.17}$$

As we will see in Chapter 4, this formula affords a direct way to obtain expressions for $\mathbf{F}$ when the positions $\mathbf{y}$ and $\mathbf{x}$ are specified in terms of coordinate systems. See the book by Steigmann for some explicit examples.

From what has been said it should be evident that, at the material point p, $d\mathbf{x}(= \mathbf{x}'ds)$ and $d\mathbf{y}(= \mathbf{y}'ds)$ are tangential to the curves $c \subset \kappa$ and $c_t \subset \kappa_t$, respectively, the latter having the parametric representation $\mathbf{y}(s,t) = \chi(\mathbf{x}(s),t)$. We call c a *material curve* , to convey the meaning that it is convected by the deformation to a curve c_t consisting of the same material points. Consider two material curves that intersect at $\mathbf{x}$, with tangents $d\mathbf{x}$ and $d\mathbf{u}$. These are transported to $d\mathbf{y}$ and $d\mathbf{z}$, respectively, where $d\mathbf{z} = \mathbf{F}d\mathbf{u}$, with $\mathbf{F} = \mathbf{F}(\mathbf{x},t)$, as in (1.17). The local state of distortion of these material curves existing at the material point p, occupying position $\mathbf{x}$ in κ, is characterized by

$$d\mathbf{y} \cdot d\mathbf{z} = \mathbf{F}d\mathbf{x} \cdot \mathbf{F}d\mathbf{u} = d\mathbf{x} \cdot \mathbf{F}^t(\mathbf{F}d\mathbf{u}) = d\mathbf{x} \cdot \mathbf{C}d\mathbf{u}, \tag{1.18}$$

where

$$\mathbf{C} = \mathbf{F}^t\mathbf{F} \tag{1.19}$$

is the right Cauchy–Green deformation tensor. Choosing the material curves to coincide, i.e., $d\mathbf{u} = d\mathbf{x}$, yields the squared stretch of a curve. Equation (1.18) then furnishes the local angle made by the tangents to two material curves after deformation.

For our purposes it will prove convenient to work with the Lagrange strain

$$\mathbf{E} = \tfrac{1}{2}(\mathbf{C} - \mathbf{I}), \tag{1.20}$$

which is in one-to-one relation to the Cauchy–Green tensor, where $\mathbf{I}$ is the referential unit tensor, defined by $\mathbf{Iv} = \mathbf{v}$ for all vectors $\mathbf{v}$ belonging to the vector space T_κ associated with κ. The latter is often called the translation space of κ, to convey the notion that it coincides with the set of all position differences that can be formed within it. Because κ resides in Euclidean space by assumption, this translation space is identical to the *tangent space* of the underlying (Euclidean) manifold. Similarly, we denote the translation (tangent) space associated with κ_t by T_{κ_t}. Thus, Euclidean spaces are effectively flat in the sense that they coincide with their tangent spaces. More will be said about this in Chapter 3.

Having discussed the bare essentials of the kinematics of deformation, we move on to the basic balance laws concerning mass and momentum.

Let $\rho(\mathbf{y},t)$ be the (positive) mass density of the body in the configuration κ_t. The mass of a subregion $\pi_t \subset \kappa_t$ is simply

$$M(\pi_t) = \int_{\pi_t} \rho dv = \int_{\pi} \rho_\kappa dV, \tag{1.21}$$

where $\pi \subset \kappa$ is the image of π_t under the inverse deformation, i.e., $\pi_t = \chi(\pi, t)$, meaning that the two regions are related by the deformation map and consist of the same set of material points; and

$$\rho_\kappa = \mathcal{J}_F \rho, \tag{1.22}$$

where $\mathcal{J}_F = |\det \mathbf{F}|$, is the referential mass density. This is simply the familiar change-of-variable formula from calculus. In this book we assume κ is occupiable, so that $\det \mathbf{F} > 0$, but this is by no means essential.

The principle of conservation of mass is the assertion that the mass of a fixed set of material points remains invariant in time. Therefore, the time derivative of $M(\pi_t)$ vanishes. Because the domain π is fixed for the material points that occupy π_t, we can pass the derivative through the right-most integral—assuming sufficient regularity of the integrand—to obtain

$$\int_\pi \dot{\rho}_\kappa dV = 0, \tag{1.23}$$

where the superposed dot is the *material derivative*, the partial time derivative holding p, and hence $\mathbf{x}$, fixed. Because π is an arbitrary subvolume of κ, assuming the integrand to be continuous we can invoke the *localization theorem*—basically the mean-value theorem for integrals—to conclude that the integrand vanishes pointwise, i.e., that $\dot{\rho}_\kappa = 0$ at every $\mathbf{x} \in \kappa$. In other words, the function $\rho_\kappa(\mathbf{x}, t) = \det \mathbf{F}(\mathbf{x}, t)\rho(\chi(\mathbf{x}, t), t)$, expressed as a function of $\mathbf{x}$ and t, is independent of t and hence a fixed function of $\mathbf{x}$. This result does not apply in the presence of diffusion, however. In this case our reasoning must be adjusted to account for the flux of mass through the boundary $\partial \pi_t$. See the book by Gurtin et al.

The balance of linear momentum is the assertion that the net force acting on the material occupying π_t is balanced by the rate of change of its momentum. Thus,

$$\int_{\partial \pi_t} \mathbf{t} da + \int_{\pi_t} \rho \mathbf{b} dv = \frac{d}{dt} \int_{\pi_t} \rho \mathbf{v} dv, \tag{1.24}$$

where $\mathbf{t}$, the *traction*, is the areal density of contact force, $\mathbf{b}$ is the body force per unit mass, and $\mathbf{v} = \dot{\mathbf{y}} = \frac{\partial}{\partial t} \chi(\mathbf{x}, t)$ is the material velocity. To reduce the right-hand side we proceed as in the reduction of the mass conservation principle. Thus, invoking conservation of mass in the form $\dot{\rho}_\kappa = 0$, we have

$$\begin{aligned} \frac{d}{dt} \int_{\pi_t} \rho \mathbf{v} dv &= \frac{d}{dt} \int_\pi \rho \mathcal{J}_F \mathbf{v} dV = \frac{d}{dt} \int_\pi \rho_\kappa \mathbf{v} dV = \int_\pi \rho_\kappa \dot{\mathbf{v}} dV \\ &= \int_\pi \rho \mathcal{J}_F \dot{\mathbf{v}} dV = \int_{\pi_t} \rho \dot{\mathbf{v}} \, dv, \end{aligned} \tag{1.25}$$

which reduces (1.24) to

$$\int_{\partial\pi_t} \mathbf{t}da + \int_{\pi_t} \rho(\mathbf{b} - \dot{\mathbf{v}})dv = \mathbf{0}. \tag{1.26}$$

Assuming the integrands to be bounded in magnitude, this may be used to establish that $\mathbf{t}$ is a function of the tangent plane to $\partial\pi_t$ at the point $\mathbf{y} \in \partial\pi_t$; equivalently, a function of the unit normal $\mathbf{n}$ to $\partial\pi_t$ at $\mathbf{y}$. See the important paper by Noll. For definiteness we take this to be the exterior unit normal. With this result in hand we may proceed via Cauchy's theorem to show that the dependence is linear, and hence that there exists a tensor field $\mathbf{T}(\mathbf{y}, t)$, the *Cauchy stress*, such that

$$\mathbf{t} = \mathbf{Tn}. \tag{1.27}$$

Substituting into (1.26) and invoking the divergence theorem, we obtain

$$\int_{\pi_t} [div\mathbf{T} + \rho(\mathbf{b} - \dot{\mathbf{v}})]dv = \mathbf{0}, \tag{1.28}$$

where $div\mathbf{A}$ is the vector field defined by

$$\mathbf{c} \cdot div\mathbf{A} = div(\mathbf{A}^t\mathbf{c}), \tag{1.29}$$

for any tensor field $\mathbf{A}(\mathbf{y}, t)$ and any *fixed* vector $\mathbf{c}$. Here $div\mathbf{w}$ is the scalar field defined by

$$div\mathbf{w} = tr(grad\mathbf{w}), \tag{1.30}$$

where tr is the trace and $grad\mathbf{w}$, the gradient of a vector field $\mathbf{w}(\mathbf{y}, t)$ with respect to $\mathbf{y}$, is the tensor field defined, as in (1.17), by

$$d\mathbf{w} = (grad\mathbf{w})d\mathbf{y}. \tag{1.31}$$

Assuming the integrand in (1.28) to be a continuous function of $\mathbf{y}$, we can localize and arrive at Cauchy's equation of motion,

$$div\mathbf{T} + \rho\mathbf{b} = \rho\dot{\mathbf{v}}, \tag{1.32}$$

holding at each $\mathbf{y} \in \kappa_t$. Generalizations to discontinuous fields will be considered in Chapter 6.

The balance of moment of momentum is the assertion that

$$\int_{\partial\pi_t} \mathbf{y} \times \mathbf{t}da + \int_{\pi_t} \mathbf{y} \times \rho\mathbf{b}dv = \tfrac{d}{dt} \int_{\pi_t} \mathbf{y} \times \rho\mathbf{v}dv, \tag{1.33}$$

where $\mathbf{y}$ is the position field relative to a fixed origin. Invoking conservation of mass, the traction formula (1.27), the linear momentum balance (1.32), and localizing as before, we arrive ultimately at the local algebraic restriction

$$\mathbf{T} = \mathbf{T}^t, \tag{1.34}$$

again at every $\mathbf{y} \in \kappa_t$. Equations (1.32) and (1.34) are often referred to as the *spatial* equations of motion.

Problem 1.1 Prove (1.34) by carrying out the steps indicated.

Equivalent *referential* forms of the equations, due essentially to Piola, may be derived with the aid of the Piola–Nanson formula

$$\mathbf{n}da = \mathbf{F}^*\boldsymbol{\nu}dA, \tag{1.35}$$

connecting the oriented area measure $\mathbf{n}da$ on $\partial\pi_t$ to its counterpart $\boldsymbol{\nu}dA$ on $\partial\pi$. Here

$$\mathbf{F}^* = \mathcal{J}_F\mathbf{F}^{-t} \tag{1.36}$$

is the *cofactor* of **F**. Thus, from (1.22), (1.26), and (1.27),

$$\int_\pi \rho_\kappa(\dot{\mathbf{v}} - \mathbf{b})dV = \int_{\pi_t} \rho(\dot{\mathbf{v}} - \mathbf{b})dv = \int_{\partial\pi_t} \mathbf{T}\mathbf{n}da = \int_{\partial\pi} \mathbf{P}\boldsymbol{\nu}dA, \tag{1.37}$$

where

$$\mathbf{P} = \mathbf{T}\mathbf{F}^* \tag{1.38}$$

is the *Piola stress*. Clearly, this provides a measure of force per unit reference area, whereas the Cauchy stress furnishes a resolution of the same force per unit area of surface after deformation. Applying the divergence theorem again, this time in the reference configuration, we have

$$\int_\pi [Div\mathbf{P} + \rho_\kappa(\mathbf{b} - \dot{\mathbf{v}})]dV = \mathbf{0}, \tag{1.39}$$

where *Div*, the divergence with respect to **x**, is defined, with obvious adjustments, in the same way that *div* was defined. Localizing as usual, we find that (1.39) is equivalent to

$$Div\mathbf{P} + \rho_\kappa \mathbf{b} = \rho_\kappa \dot{\mathbf{v}}, \tag{1.40}$$

holding at each $\mathbf{x} \in \kappa$, this formulation having the convenient feature that the function $\rho_\kappa(\mathbf{x})$ is known a priori, whereas the symmetry condition (1.34) is equivalent to

$$\mathbf{P}\mathbf{F}^t = \mathbf{F}\mathbf{P}^t. \tag{1.41}$$

To express the latter condition in a more convenient form we introduce the *Piola–Kirchhoff stress* **S**, defined by

$$\mathbf{P} = \mathbf{F}\mathbf{S}. \tag{1.42}$$

Then $\mathbf{P}\mathbf{F}^t = \mathbf{F}\mathbf{S}\mathbf{F}^t$ and the invertibility of **F** implies that (1.41) is equivalent to the symmetry

$$\mathbf{S} = \mathbf{S}^t. \tag{1.43}$$

Before concluding these preliminaries we pause to state the mechanical energy balance,

$$\mathcal{P}(\pi, t) = \mathcal{S}(\pi, t) + \tfrac{d}{dt}\mathcal{K}(\pi, t), \tag{1.44}$$

where

$$\mathcal{K}(\pi, t) = \tfrac{1}{2}\int_\pi \rho_\kappa \mathbf{v} \cdot \mathbf{v}dV \tag{1.45}$$

is the kinetic energy of the material occupying π,

$$\mathcal{S}(\pi, t) = \int_\pi \mathbf{P} \cdot \dot{\mathbf{F}}dV \tag{1.46}$$

is the *stress power*, in which $\mathbf{A} \cdot \mathbf{B} = tr(\mathbf{A}\mathbf{B}^t)$ is the *inner product* of tensors **A** and **B**, and

$$\mathcal{P}(\pi, t) = \int_{\partial\pi} \mathbf{p} \cdot \mathbf{v}dA + \int_\pi \rho_\kappa \mathbf{b} \cdot \mathbf{v}dV \tag{1.47}$$

is the power of the forces acting on the material in π, in which

$$\mathbf{p} = \mathbf{P}\boldsymbol{\nu} \tag{1.48}$$

is the Piola traction, related to the Cauchy traction $\mathbf{t}$ by

$$\int_{\partial\pi} \mathbf{p}dA = \int_{\partial\pi_t} \mathbf{t}da. \tag{1.49}$$

The mechanical energy balance will play a central role in our development.

Problem 1.2 Derive (1.44) from the momentum balance (1.40). Hint: Dot multiply (1.40) by the material velocity $\mathbf{v}$. Show that $\mathbf{v} \cdot Div\mathbf{P} = Div(\mathbf{P}^t\mathbf{v}) - \mathbf{P} \cdot \nabla\mathbf{v}$, where ∇ is the gradient with respect to $\mathbf{x}$, and that $\nabla\mathbf{v} = \dot{\mathbf{F}}$. Integrate over $\pi \subset \kappa$ and invoke the divergence theorem.

With reference to the problem, note that if we dot multiply (1.40) by an arbitrary vector field $\mathbf{u}$ instead of the material velocity $\mathbf{v}$, we arrive at the integral statement

$$\int_{\partial\pi} \mathbf{p} \cdot \mathbf{u}dA + \int_{\pi} \rho_\kappa(\mathbf{b} - \dot{\mathbf{v}}) \cdot \mathbf{u}dV = \int_{\pi} \mathbf{P} \cdot \nabla\mathbf{u}dV \tag{1.50}$$

in place of (1.44). It is usual, though no doubt unwise from the pedagogical point of view, to call $\mathbf{u}$ a *virtual* velocity field, or worse, a virtual *displacement*, to emphasize the fact that it has nothing whatever to do with the actual material velocity. This is the *weak form* of the equation of motion, so named because it requires a weaker degree of regularity than the local, or *strong*, form.

Although we have obtained it as a necessary condition for (1.40), it is also sufficient. To see this we simply start with (1.50), write $\mathbf{P} \cdot \nabla\mathbf{u} = Div(\mathbf{P}^t\mathbf{u}) - \mathbf{u} \cdot Div\mathbf{P}$, and apply the divergence theorem, reaching

$$\int_{\partial\pi} (\mathbf{p} - \mathbf{P}\boldsymbol{\nu}) \cdot \mathbf{u}dA + \int_{\pi} \{Div\mathbf{P} + \rho_\kappa(\mathbf{b} - \dot{\mathbf{v}})\} \cdot \mathbf{u}dV = \mathbf{0}. \tag{1.51}$$

As $\mathbf{u}$ is arbitrary, we choose

$$\mathbf{u} = f(\mathbf{x})^2\{Div\mathbf{P} + \rho_\kappa(\mathbf{b} - \dot{\mathbf{v}})\}, \tag{1.52}$$

where $f(\mathbf{x})$ is any function that vanishes on $\partial\pi$, thereby reducing (1.51) to

$$\int_{\pi} f^2 \left|Div\mathbf{P} + \rho_\kappa(\mathbf{b} - \dot{\mathbf{v}})\right|^2 dV = 0, \tag{1.53}$$

and hence requiring that (1.40) hold locally in π, leaving

$$\int_{\partial\pi} (\mathbf{p} - \mathbf{P}\boldsymbol{\nu}) \cdot \mathbf{u}dA = 0 \tag{1.54}$$

as the remaining content of (1.51). Choosing $\mathbf{u} = \mathbf{p} - \mathbf{PN}$ on $\partial\pi$ then yields

$$\int_{\partial\pi} |\mathbf{p} - \mathbf{P}\boldsymbol{\nu}|^2 \, dA = 0, \tag{1.55}$$

which in turn requires that (1.48) hold locally, at each point of $\partial\pi$.

More often this procedure is invoked with π replaced by κ. In this case position $\mathbf{y} = \boldsymbol{\chi}(\mathbf{x}, t)$ is typically assigned as a fixed function $\boldsymbol{\phi}(\mathbf{x})$, say, on a part $\partial\kappa_y$ of the boundary $\partial\kappa$, implying that the actual velocity $\mathbf{v}$ vanishes there. We then stipulate that $\mathbf{u}$ should also vanish on $\partial\kappa_y$, as a condition of a so-called *kinematically admissible* virtual velocity field. The argument leading to (1.53), with π replaced by κ, remains valid, but in place of (1.54) we now have

$$\int_{\partial\kappa_p} (\mathbf{p} - \mathbf{P}\boldsymbol{\nu}) \cdot \mathbf{u} dA = 0, \tag{1.56}$$

where $\partial\kappa_p = \partial\kappa \setminus \partial\kappa_y$. Choosing

$$\mathbf{u} = g(\mathbf{x})^2 (\mathbf{p} - \mathbf{P}\boldsymbol{\nu}), \tag{1.57}$$

where $g(\mathbf{x})$ is any function that vanishes on the curve(s) $\partial(\partial\kappa_p) = \partial(\partial\kappa_y)$ in accordance with kinematic admissibility, we then have

$$\int_{\partial\kappa_p} g^2 \, |\mathbf{p} - \mathbf{P}\boldsymbol{\nu}|^2 \, dA = 0, \tag{1.58}$$

which requires that (1.48) hold pointwise on $\partial\kappa_p$.

References

Batchelor, G. K. (Ed.) (1958). *The Scientific Papers of Sir Geoffrey Ingram Taylor*, Vol. 1: *Mechanics of Solids*. Cambridge University Press, Cambridge, UK.

Chadwick, P. (1976). *Continuum Mechanics: Concise Theory and Problems*. Dover, New York.

Epstein, M., and Elżanowski, M (2007). *Material Inhomogeneities and Their Evolution*. Springer, Berlin.

Fleming, W. (1977). *Functions of Several Variables*. Springer, Berlin.

Gurtin, M. E. (1981). *An Introduction to Continuum Mechanics*. Academic Press, Orlando.

Gurtin, M. E., Fried, E., and Anand, L. (2010). *The Mechanics and Thermodynamics of Continua*. Cambridge University Press, Cambridge, UK.

Havner, K. S. (1992). *Finite Plastic Deformation of Crystalline Solids*. Cambridge University Press, Cambridge, UK.

Hill, R. (1950). *The Mathematical Theory of Plasticity*. Oxford University Press, Oxford.

Liu, I-Shih. (2002). *Continuum Mechanics*. Springer, Berlin.

Kovács, I., and Zsoldos, L. (1973). *Dislocations and Plastic Deformation*. Pergamon Press, Oxford.

Maugin, G. A. (1992). *The Thermomechanics of Plasticity and Fracture.* Cambridge University Press, Cambridge, UK.

Noll, W. (1974). The foundations of classical mechanics in the light of recent advances in continuum mechanics. Reprinted in Truesdell, C. (Ed.), *The Foundations of Mechanics and Thermodynamics,* pp. 32–47 Springer, Berlin.

Steigmann, D. J. (2017). *Finite Elasticity Theory.* Oxford University Press, Oxford.

Teodosiu, C. (Ed) (1997). *Large Plastic Deformation of Crystalline Aggregates.* CISM Courses and Lectures, No. 376. Springer, Vienna.

2

Brief resumé of nonlinear elasticity theory

Familiarity with the basic elements of nonlinear elasticity theory is essential to a proper understanding of virtually the entire range of topics comprising solid mechanics in general, and plasticity theory in particular. Accordingly we devote the present chapter to a brief survey of those aspects of nonlinear elasticity that will prove central to our later work.

2.1 Stress and strain energy

Following Noll's landmark 1958 paper, we define *elasticity* to mean that the value of the Cauchy stress $\mathbf{T}(p, t)$ existing at the material point $p \in B$ at time t is determined by the present value of the deformation function $\boldsymbol{\chi}(\mathbf{x}', t)$ for $\mathbf{x}' \in N_\kappa(\mathbf{x})$, an arbitrary, and hence arbitrarily small, neighborhood of the place $\mathbf{x}$ occupied by p in reference configuration κ. Our assumption of differentiability of the deformation—see (1.10)—implies that the deformations influencing this stress are approximated by

$$\boldsymbol{\chi}(\mathbf{x}', t) = \boldsymbol{\chi}(\mathbf{x}, t) + \mathbf{F}(\mathbf{x}, t)(\mathbf{x}' - \mathbf{x}) + \mathbf{o}(|\mathbf{x}' - \mathbf{x}|), \tag{2.1}$$

and are therefore determined, at linear-order accuracy, by $\boldsymbol{\chi}(\mathbf{x}, t)$ and $\mathbf{F}(\mathbf{x}, t)$.

The requirement of frame invariance satisfied by all sensible constitutive functions implies that the constitutive equation giving the stress is, among other things, *translation invariant* in the sense that it remains invariant under constant time translations and perturbations of the position $\mathbf{y}$ currently occupied by a material point. It is therefore not explicitly dependent on t or $\boldsymbol{\chi}(\mathbf{x}, t)$. Accordingly, at leading order the stress is determined by $\mathbf{F}(\mathbf{x}, t)$. Termination at this order leads to Noll's *simple material* model of elasticity, according to which

$$\mathbf{T}(p, t) = \mathcal{T}_\kappa(\mathbf{F}(\mathbf{x}, t); \mathbf{x}) \tag{2.2}$$

for some *constitutive function* $\mathcal{T}_\kappa$, the subscript identifying the reference configuration relative to which the deformation gradient is computed. It is best to make this explicit to

A Course on Plasticity Theory. David J. Steigmann, Oxford University Press. © David J. Steigmann (2022).
DOI: 10.1093/oso/9780192883155.003.0002

avoid confusion when choosing alternative reference configurations, as we shall do later. Of course it is possible to keep further terms in the expansion (2.1) and to contemplate their influence on constitutive equations for so-called *materials of higher grade.* While these are important and useful, they are of limited relevance to the current state of the art in plasticity theory. Given the constitutive function $\mathcal{T}_\kappa$ for the Cauchy stress, that for the Piola stress, for example, relative to the same reference configuration, follows easily from (1.38):

$$\mathbf{P}(p, t) = \mathcal{P}_\kappa(\mathbf{F}(\mathbf{x}, t); \mathbf{x}) = \mathcal{T}_\kappa(\mathbf{F}(\mathbf{x}, t); \mathbf{x})\mathbf{F}^*(\mathbf{x}, t). \tag{2.3}$$

Frame invariance imposes further restrictions on these constitutive functions, but to save time we defer these and proceed instead to a further specialization of the purely mechanical theory. This is the so-called *work inequality*, which posits that it is necessary to perform non-negative work on a sample of material to cause it to undergo a hypothetical cyclic process; that is, a process in which the deformation and velocity fields at the start and end of a process, occurring in a time interval $[t_1, t_2]$, coincide at every material point, i.e.,

$$\chi(\mathbf{x}, t_1) = \chi(\mathbf{x}, t_2) \quad \text{and} \quad \dot{\chi}(\mathbf{x}, t_1) = \dot{\chi}(\mathbf{x}, t_2). \tag{2.4}$$

Taking gradients, these imply that

$$\mathbf{F}(\mathbf{x}, t_1) = \mathbf{F}(\mathbf{x}, t_2) \quad \text{and} \quad \dot{\mathbf{F}}(\mathbf{x}, t_1) = \dot{\mathbf{F}}(\mathbf{x}, t_2). \tag{2.5}$$

This assumption is related to the hypothesis known more popularly as the non-existence of perpetual motion machines. However, given that all material points of the sample are involved, the creation of a cyclic process is no small feat from the experimental point of view. Moreover, while this assumption has a thermodynamical flavor, it is not in fact a consequence of any principle of thermodynamics. It is nevertheless realistic, and in accord with everyday experience.

Proceeding, we impose the work inequality on an arbitrary subvolume $\pi \subset \kappa$, invoke (1.44), and, noting that the kinetic energies at the start and end of the cycle coincide, conclude that

$$\int_{t_1}^{t_2} \mathcal{S}(\pi, t)dt \geq 0, \tag{2.6}$$

where $\mathcal{S}$ is the stress power defined in (1.46). Assuming the integrand therein to be continuous in $[t_1, t_2] \times \pi$, Fubini's theorem (see Fleming's book) ensures that we can interchange the order of integration, and hence that

$$\int_\pi \int_{t_1}^{t_2} \mathbf{P} \cdot \dot{\mathbf{F}} dt \geq 0 \quad \text{for all} \quad \pi \subset \kappa. \tag{2.7}$$

Localizing as usual, we have

$$\int_{t_1}^{t_2} \mathbf{P} \cdot \dot{\mathbf{F}} dt \geq 0, \quad \text{at each} \quad \mathbf{x} \in \kappa, \tag{2.8}$$

and hence, in the case of elasticity,

$$\oint \mathcal{P}_\kappa(\mathbf{F}) \cdot d\mathbf{F} \geq 0, \tag{2.9}$$

in which the integral is taken round a smooth curve in **F**-space, in accordance with the definition of a cyclic process, and the parametric dependence on **x**, having no bearing on the discussion, has been suppressed. By considering the reversal of this cyclic process—itself cyclic—it is possible to show—see the books by Gurtin and Steigmann, for example—that this statement is satisfied as an equality for all smooth closed curves. This, in turn, is necessary and sufficient for the existence of a *strain-energy* function $\Psi(\mathbf{F}; \mathbf{x})$, say, such that

$$\mathbf{P} = \Psi_{\mathbf{F}}, \tag{2.10}$$

the gradient of Ψ with respect to **F**. Given the function Ψ, this gradient is easily computed via the chain rule by writing $d\Psi$ as a linear form in $d\mathbf{F}$ and reading off the coefficient tensor, i.e.,

$$d\Psi = \Psi_{\mathbf{F}} \cdot d\mathbf{F}. \tag{2.11}$$

We thus have

$$\mathcal{P}_\kappa(\mathbf{F}; \mathbf{x}) = \Psi_{\mathbf{F}}(\mathbf{F}; \mathbf{x}), \tag{2.12}$$

and hence the rather remarkable conclusion that the stress is determined by a single scalar-valued function. We then refer, somewhat dramatically, to a *hyperelastic material*. The strain-energy function has as arguments variables defined on κ, which should therefore be appended as a subscript. In an effort to minimize clutter we normally refrain from doing this when the context is clear.

On combining (2.2), (2.3), and (2.10), the symmetry condition (1.34) satisfied by the Cauchy stress may be expressed in terms of the strain-energy function as

$$(\Psi_{\mathbf{F}})\mathbf{F}^t = \mathbf{F}(\Psi_{\mathbf{F}})^t, \tag{2.13}$$

which in turn is equivalent to the statement

$$(\Psi_{\mathbf{F}})\mathbf{F}^t \cdot \mathbf{\Omega} = 0 \tag{2.14}$$

for arbitrary skew $\mathbf{\Omega}$; that is, for all $\mathbf{\Omega}$ such that $\mathbf{\Omega}^t = -\mathbf{\Omega}$. The identity

$$\mathbf{AB} \cdot \mathbf{D} = \mathbf{A} \cdot \mathbf{DB}^t \tag{2.15}$$

allows us to rewrite this as

$$\Psi_{\mathbf{F}} \cdot \mathbf{\Omega F} = 0. \tag{2.16}$$

Our intention is to characterize all functions Ψ that satisfy this requirement. If we succeed, then we can replace the symmetry of the Cauchy stress once and for all by the general form of Ψ to be derived. Before undertaking this task, however, a brief digression is in order.

Consider the initial-value problem

$$\dot{\mathbf{Q}} = \mathbf{\Omega Q}; \quad \mathbf{Q}(0) = \mathbf{i}, \tag{2.17}$$

where $\mathbf{Q}(u)$ is a one-parameter family of tensors, the superposed dot stands for the derivative with respect to u, $\mathbf{\Omega}$ is a fixed, but arbitrary, skew tensor, and $\mathbf{i}$ is the spatial identity tensor. We want to show that $\mathbf{Q}(u)$ is a rotation tensor for all values of the parameter u.

To this end we define

$$\mathbf{Z}(u) = \mathbf{Q}(u)\mathbf{Q}(u)^t, \tag{2.18}$$

and find, with (2.17), that this satisfies the initial-value problem

$$\dot{\mathbf{Z}} = \mathbf{\Omega Z} - \mathbf{Z\Omega}; \quad \mathbf{Z}(0) = \mathbf{i}. \tag{2.19}$$

Clearly $\mathbf{Z}(u) = \mathbf{i}$ is a solution, and the uniqueness theorem for solutions to ordinary differential equations implies that there is no other. Thus, $\mathbf{Q}(u)$ is orthogonal and hence invertible. To establish that it is a rotation, we need to show that its determinant equals unity. This follows from the the fact that the cofactor of a tensor is the derivative of its determinant, i.e.,

$$\dot{\mathcal{J}}_Q = \mathbf{Q}^* \cdot \dot{\mathbf{Q}} = \mathcal{J}_Q \mathbf{Q}^{-t} \cdot \dot{\mathbf{Q}} = \mathcal{J}_Q tr(\dot{\mathbf{Q}}\mathbf{Q}^{-1}) = \mathcal{J}_Q tr\mathbf{\Omega}, \tag{2.20}$$

which vanishes because $\mathbf{\Omega}$ is skew, implying that $\mathcal{J}_{Q(u)} = \mathcal{J}_{Q(0)} = 1$, as claimed.

Returning to the task at hand, consider the one-parameter family

$$\mathbf{F}(u) = \mathbf{Q}(u)\mathbf{F}_0; \quad \mathbf{F}_0 = \mathbf{F}(0), \tag{2.21}$$

with $\mathbf{Q}(u)$ as in the foregoing. Because this pertains to a single (but arbitrary) material point, the fact that $\mathbf{F}$ is the gradient of a deformation does not impose any restriction on its values beyond $\mathcal{J}_{F_0} > 0$, assuming the reference configuration to be occupiable. We then have

$$\dot{\mathbf{F}} = \dot{\mathbf{Q}}\mathbf{F}_0 = \dot{\mathbf{Q}}\mathbf{Q}^t\mathbf{F} = \mathbf{\Omega F}. \tag{2.22}$$

Accordingly, for this choice (2.16) reduces to

$$0 = \Psi_{\mathbf{F}} \cdot \dot{\mathbf{F}} = \dot{\Psi}, \tag{2.23}$$

or simply that $\Psi(\mathbf{F}(u)) = \Psi(\mathbf{F}_0)$, where we have again suppressed the passive argument $\mathbf{x}$. Dropping the subscript and reinstating this argument, we conclude that the symmetry of the Cauchy stress implies the invariance of the energy under superposed rotations, i.e.,

$$\Psi(\mathbf{QF}; \mathbf{x}) = \Psi(\mathbf{F}; \mathbf{x}) \tag{2.24}$$

for any rotation $\mathbf{Q}$. For $\mathbf{Q}$ spatially uniform, this is precisely the condition of frame invariance of the energy, which implies that it is insensitive to arbitrary rigid-body motions

$$\chi(\mathbf{x}, t) \to \mathbf{Q}(t)\chi(\mathbf{x}, t) + \mathbf{c}(t) \tag{2.25}$$

superposed on the deformation $\chi(\mathbf{x}, t)$, where $\mathbf{c}$ is an arbitrary vector. See the treatment of frame invariance in the important paper by Murdoch and its adaptation to elasticity in the book by Steigmann.

We have shown that (2.24) is necessary for (1.34). It is also sufficient. To demonstrate this, recall the definition (1.19) of the Cauchy–Green deformation tensor. Consider two deformation gradients $\mathbf{F}$ and $\tilde{\mathbf{F}}$ and let $\tilde{\mathbf{Q}} = \tilde{\mathbf{F}}\mathbf{F}^{-1}$. Let $\mathbf{C}$ and $\tilde{\mathbf{C}}$ respectively be the Cauchy–Green tensors formed from $\mathbf{F}$ and $\tilde{\mathbf{F}}$.

Problem 2.1 If $\tilde{\mathbf{C}} = \mathbf{C}$, show that $\tilde{\mathbf{Q}}^t\tilde{\mathbf{Q}} = \mathbf{i}$, so that $\tilde{\mathbf{Q}}$ is orthogonal. Conversely, show that if $\tilde{\mathbf{Q}}$ is orthogonal, then $\tilde{\mathbf{C}} = \mathbf{C}$.

Suppose that $\Psi(\tilde{\mathbf{F}}; \mathbf{x}) = \Psi(\mathbf{F}; \mathbf{x})$ whenever $\tilde{\mathbf{F}} = \tilde{\mathbf{Q}}\mathbf{F}$, with $\tilde{\mathbf{Q}}$ a rotation and hence orthogonal. This is simply a restatement of (2.24). Then, from the problem, it follows that $\Psi(\tilde{\mathbf{F}}; \mathbf{x}) = \Psi(\mathbf{F}; \mathbf{x})$ whenever $\tilde{\mathbf{C}} = \mathbf{C}$. This means that Ψ depends on $\mathbf{F}$ through the induced Cauchy–Green tensor, and hence that

$$\Psi(\mathbf{F}; \mathbf{x}) = \hat{\Psi}(\mathbf{C}; \mathbf{x}) \tag{2.26}$$

for some function $\hat{\Psi}$. Equivalently,

$$\Psi(\mathbf{F}; \mathbf{x}) = U(\mathbf{E}; \mathbf{x}), \tag{2.27}$$

where, from (1.20), $U(\mathbf{E}; \mathbf{x}) = \hat{\Psi}(2\mathbf{E} + \mathbf{I}; \mathbf{x})$.

Problem 2.2 We have shown that (2.24) implies (2.26). Show that (2.26) implies (2.24), and hence that the two statements are equivalent.

Consider a one-parameter $\mathbf{F}(u)$ of deformation gradients. This induces a one-parameter family of strains, given by

$$\mathbf{E}(u) = \tfrac{1}{2}[\mathbf{F}(u)^t\mathbf{F}(u) - \mathbf{I}]. \tag{2.28}$$

Substituting these into (2.27) and differentiating, we have

$$\Psi_{\mathbf{F}} \cdot \dot{\mathbf{F}} = U_{\mathbf{E}} \cdot \dot{\mathbf{E}}, \tag{2.29}$$

where $U_{\mathbf{E}}$ is the gradient of the function $U(\cdot; \mathbf{x})$, and

$$\dot{\mathbf{E}} = \tfrac{1}{2}(\dot{\mathbf{F}}^t\mathbf{F} + \mathbf{F}^t\dot{\mathbf{F}}) = Sym(\mathbf{F}^t\dot{\mathbf{F}}), \tag{2.30}$$

in which $Sym(\cdot)$ is the symmetric part of its tensor argument. Because $U_{\mathbf{E}}$ occurs in an inner product with a symmetric tensor, and because the set of symmetric tensors constitutes a linear space (i.e., a 6-dimensional vector space), we may assume, without loss of generality, that $U_{\mathbf{E}}$ is a symmetric tensor. For, the skew part, if non-zero, would be annihilated by the inner product. We thus have

$$\Psi_{\mathbf{F}} \cdot \dot{\mathbf{F}} = U_{\mathbf{E}} \cdot Sym(\mathbf{F}^t\dot{\mathbf{F}}) = U_{\mathbf{E}} \cdot \mathbf{F}^t\dot{\mathbf{F}} = \mathbf{F}(U_{\mathbf{E}}) \cdot \dot{\mathbf{F}}, \tag{2.31}$$

after making use of the identity $\mathbf{A} \cdot \mathbf{BD} = \mathbf{B}^t\mathbf{A} \cdot \mathbf{D}$.

Problem 2.3 Prove this identity and the identity (2.15).

We thus conclude that

$$\Psi_{\mathbf{F}} = \mathbf{F}(U_{\mathbf{E}}). \tag{2.32}$$

Because the left-hand side is just the Piola stress, and in view of the definition (1.42), it follows that the Piola–Kirchhoff stress is given by

$$\mathbf{S} = U_{\mathbf{E}}. \tag{2.33}$$

This is automatically symmetric, of course, and (1.38) then implies that the Cauchy stress is too. This concludes our proof that (2.24) is sufficient for the symmetry of the Cauchy stress. With the previous proof of necessity, and the fact that (2.24) is necessary and sufficient for (2.27), we conclude that (2.27) is *equivalent* to the symmetry of the Cauchy stress for elastic materials that satisfy the work inequality.

Problem 2.4 Show that $\mathcal{P}_\mathcal{K}(\mathbf{QF};\mathbf{x}) = \mathbf{Q}\mathcal{P}_\mathcal{K}(\mathbf{F};\mathbf{x})$ and $\mathcal{T}_\mathcal{K}(\mathbf{QF};\mathbf{x}) = \mathbf{Q}\mathcal{T}_\mathcal{K}(\mathbf{F};\mathbf{x})\mathbf{Q}^t$.

2.2 Conservative problems and potential energy

Hyperelasticity affords a rich framework for the analysis of motion, which will be familiar to those versed in analytical dynamics. From (1.46) and (2.10), it yields the stress power in π as a time derivative,

$$\mathcal{S}(\pi, t) = \tfrac{d}{dt}\mathcal{U}(\pi, t), \tag{2.34}$$

where

$$\mathcal{U}(\pi, t) = \int_\pi \Psi dV \tag{2.35}$$

is the total strain energy in π. We may write this alternatively as

$$\mathcal{U} = \int_{\pi_t} \psi dv, \tag{2.36}$$

where $\psi(\mathbf{F};\mathbf{x})$, given by

$$\Psi = \mathcal{J}_F \psi, \tag{2.37}$$

is the strain energy per unit current volume.

If the forces acting on the material in κ are such that the power they generate is derivable from a *load potential* $\mathcal{L}$, i.e., if there exists $\mathcal{L}(\kappa, t)$ such that

$$\mathcal{P}(\kappa, t) = \tfrac{d}{dt}\mathcal{L}(\kappa, t), \tag{2.38}$$

then the mechanical energy balance (1.44), applied with $\pi = \kappa$, becomes

$$\tfrac{d}{dt}E(\kappa, t) = 0, \tag{2.39}$$

where

$$E(\kappa, t) = \mathcal{K}(\kappa, t) + \mathcal{E}[\chi], \tag{2.40}$$

in which

$$\mathcal{E} = \mathcal{U}(\kappa, t) - \mathcal{L}(\kappa, t) \tag{2.41}$$

is the total mechanical energy. Here $\mathcal{E}$ is the total *potential* energy of the configuration $\chi(\kappa, t)$, i.e., the energy of elasticity and the forces together. Thus, the total energy is conserved, and such loads are said to be *conservative*. Of course elasticity theory, and hyperelasticity in particular, is perfectly sensible for non-conservative loads too, but these do not give rise to any conservation law of this kind. Nor, in fact, does the response of any real material, even if the loading can be associated with a load potential. In reality, (2.39) is replaced by the inequality

$$\tfrac{d}{dt}E(\kappa, t) \leq 0, \tag{2.42}$$

due to effects such as heat conduction, viscosity, friction, and indeed plasticity, as we shall see later, that dissipate energy and which are not taken into account in pure elasticity theory.

Nevertheless, the notion of energy is extremely useful if interpreted appropriately. For example, consider a motion $\chi(\mathbf{x}, t)$ and an instant t_0 such that

$$\chi(\mathbf{x}, t_0) = \chi_0(\mathbf{x}) \quad \text{and} \quad \dot{\chi}(\mathbf{x}, t)_{|t_0} = \mathbf{0}. \tag{2.43}$$

Suppose the motion to be such that

$$\lim_{t\to\infty} \chi(\mathbf{x}, t) = \chi_\infty(\mathbf{x}) \quad \text{and} \quad \lim_{t\to\infty} \dot{\chi}(\mathbf{x}, t) = \mathbf{0}. \tag{2.44}$$

This limiting state, if it exists, is said to be *asymptotically stable* relative to the state existing at time t_0. It then follows from (2.40) and (2.42) that asymptotically stable deformations minimize the potential energy, i.e.,

$$\mathcal{E}[\chi_\infty] \leq \mathcal{E}[\chi_0]. \tag{2.45}$$

Thus, the problem of finding stable deformations is reduced to the problem of minimizing the potential energy, presuming, of course, that such an energy exists, i.e., that the material is hyperelastic and the loading is conservative.

Problem 2.5 (a) Consider the case of *dead loading* in which the Piola traction and body force are assigned as fixed functions of $\mathbf{x}$, independent of time, on $\partial\kappa_p = \partial\kappa \setminus \partial\kappa_y$ and in κ, respectively. Show that this loading is conservative, and that the potential energy is

$$\mathcal{E}[\chi] = \int_\kappa (\Psi - \rho_\kappa \mathbf{b} \cdot \chi) dV - \int_{\partial\kappa_p} \mathbf{p} \cdot \chi dA,$$

apart from a constant. The constant, which has not been made explicit, is unimportant. Why is that?

(b) State conditions under which a pressure load $\mathbf{t} = -p\mathbf{n}$ is conservative, where $\mathbf{t}$ is the Cauchy traction, p is the pressure, and $\mathbf{n}$ is the exterior unit normal to the boundary $\partial\kappa_t$. Obtain expressions for the associated load potentials.

Problem 2.6 Consider a viscoelastic material model in which the Piola stress is given by

$$\mathbf{P} = \Psi_{\mathbf{F}} + \mu \mathbf{F}\dot{\mathbf{E}},$$

where μ is the viscosity of the material and $\dot{\mathbf{E}}$ is the material derivative of the Lagrange strain. This model is too simple to furnish a quantitative description of the actual viscoelastic response of most materials but nevertheless exhibits the correct qualitative behavior.

Define the *dissipation* $\mathcal{D}$ as the difference between the power of the applied loads and the rate of change of the sum of the strain and kinetic energies, i.e.,

$$\mathcal{D} = \mathcal{P} - \tfrac{d}{dt}(\mathcal{U} + \mathcal{K}).$$

Thus, from (1.44) and (2.34), $\mathcal{D}$ vanishes in the case of purely elastic response ($\mu = 0$).

(a) Show that $\mathcal{D} \geq 0$ if the viscosity is positive and that $\mathcal{D}$ then vanishes *if and only if* $\dot{\mathbf{E}} = \mathbf{0}$ at all material points.

(b) In the case of conservative loading show that the total energy is dissipated, i.e.,

$$\tfrac{d}{dt}E \leq 0,$$

if the viscosity is positive.

Let $\chi(\mathbf{x}; u)$ be a one-parameter family of deformations, and denote $\chi(\mathbf{x}; 0)$ simply by $\chi(\mathbf{x})$. We define kinematic admissibility of this family to mean that $\chi(\mathbf{x}; u)$ is as smooth as required by the analysis to follow and that it equals the fixed, u-independent function $\boldsymbol{\phi}(\mathbf{x})$, say, on $\partial\kappa_y$. We are interested in characterizing energy-minimizing deformations $\chi(\mathbf{x})$; namely, those functions such that

$$\mathcal{E}[\chi(\mathbf{x})] \leq \mathcal{E}[\chi(\mathbf{x}; u)]. \tag{2.46}$$

To this end we define

$$f(u) = \mathcal{E}[\chi(\mathbf{x}; u)]. \tag{2.47}$$

Then, for (2.46) it is necessary that

$$\dot{f} = 0 \quad \text{and} \quad \ddot{f} \geq 0, \tag{2.48}$$

where

$$\dot{f} = f'(u)_{|u=0} \quad \text{and} \quad \ddot{f} = f''(u)_{|u=0} \tag{2.49}$$

are the first and second variations of the potential energy, respectively.

In the case of dead loading, for example,

$$f(u) = \int_{\kappa} \{\Psi(\mathbf{F}(\mathbf{x};u);\mathbf{x}) - \rho_{\kappa}\mathbf{b}\cdot\chi(\mathbf{x};u)\}dV - \int_{\partial\kappa_p} \mathbf{p}\cdot\chi(\mathbf{x};u)dA, \tag{2.50}$$

where $\mathbf{F}(\mathbf{x};u) = \nabla\chi(\mathbf{x};u)$, and the first and second variations are

$$\dot{f} = \int_{\kappa} (\mathbf{P}\cdot\nabla\mathbf{u} - \rho_{\kappa}\mathbf{b}\cdot\mathbf{u})dV - \int_{\partial\kappa_p} \mathbf{p}\cdot\mathbf{u}dA \tag{2.51}$$

and

$$\begin{aligned}\ddot{f} \;=\; & \int_{\kappa} \mathcal{M}[\nabla\mathbf{u}]\cdot\nabla\mathbf{u}dV \\ & + \int_{\kappa} (\mathbf{P}\cdot\nabla\mathbf{v} - \rho_{\kappa}\mathbf{b}\cdot\mathbf{v})dV - \int_{\partial\kappa_p} \mathbf{p}\cdot\mathbf{v}dA,\end{aligned} \tag{2.52}$$

respectively, where

$$\mathbf{u}(\mathbf{x}) = \frac{\partial}{\partial u}\chi(\mathbf{x};u)_{|u=0} \quad \text{and} \quad \mathbf{v}(\mathbf{x}) = \frac{\partial^2}{\partial u^2}\chi(\mathbf{x};u)_{|u=0} \tag{2.53}$$

are the virtual velocity and acceleration. These vanish identically on $\partial\kappa_y$. Further,

$$\mathbf{P} = \Psi_{\mathbf{F}}(\nabla\chi(\mathbf{x});\mathbf{x}) \tag{2.54}$$

is the Piola stress associated with the minimizing deformation $\chi(\mathbf{x})$, and

$$\mathcal{M} = \Psi_{\mathbf{FF}}(\nabla\chi(\mathbf{x});\mathbf{x}) \tag{2.55}$$

is the fourth-order tensor representing the second derivative of the strain-energy function, again evaluated at the minimizing deformation. This presumes, of course, that the strain energy is twice differentiable with respect to the deformation gradient, this in turn implying the *major symmetry* condition

$$\mathcal{M} = \mathcal{M}^t, \tag{2.56}$$

where the transpose $\mathcal{M}^t$ is defined by

$$\mathbf{A}\cdot\mathcal{M}^t[\mathbf{B}] = \mathbf{B}\cdot\mathcal{M}[\mathbf{A}] \tag{2.57}$$

for all second-order tensors $\mathbf{A}, \mathbf{B}$; and, here and in (2.52), the square bracket notation is used to denote the second-order tensor resulting from the linear action of a fourth-order tensor on a second-order tensor. In terms of Cartesian components, for example, $(\mathcal{M}[\mathbf{A}])_{ij} = \mathcal{M}_{ijkl}A_{kl}$, in which a double sum on k and l from one to three is implied.

Recalling that **u** vanishes on $\partial\kappa_p$, we use the divergence theorem in (2.51) to reduce $(2.48)_1$, with (2.51), to

$$\int_{\partial\kappa_p} (\mathbf{P}\boldsymbol{\nu} - \mathbf{p}) \cdot \mathbf{u}dA - \int_{\kappa} (Div\mathbf{P} + \rho_{\kappa}\mathbf{b}) \cdot \mathbf{u}dV = 0, \tag{2.58}$$

and proceed as in Chapter 1 to conclude that

$$Div\mathbf{P} + \rho_{\kappa}\mathbf{b} = \mathbf{0} \quad \text{in} \quad \kappa, \quad \text{and} \quad \mathbf{p} = \mathbf{P}\boldsymbol{\nu} \quad \text{on} \quad \partial\kappa_p, \tag{2.59}$$

and hence that energy minimizers are *equilibrium* deformations. Further, as **u** is an arbitrary vector field that vanishes on $\partial\kappa_y$, and as **v** is such a field, the second line in (2.52) vanishes by (2.58), and $(2.48)_2$ reduces to

$$\int_{\kappa} \mathcal{M}(\mathbf{F}(\mathbf{x}); \mathbf{x})[\nabla\mathbf{u}] \cdot \nabla\mathbf{u}dV \geq 0, \tag{2.60}$$

for all such **u**.

2.3 The Legendre–Hadamard inequality

We proceed to show that the deformation $\chi(\mathbf{x})$, with gradient $\mathbf{F}(\mathbf{x}) = \nabla\chi$, must then be such that the *Legendre–Hadamard* inequality

$$\mathbf{a} \otimes \mathbf{n} \cdot \mathcal{M}(\mathbf{F}; \mathbf{x})[\mathbf{a} \otimes \mathbf{n}] \geq 0 \tag{2.61}$$

is satisfied at every point $\mathbf{x} \in \kappa$ and for all vectors **a** and **n**.

Proof. Consider

$$\mathbf{u}(\mathbf{x}) = \epsilon\boldsymbol{\xi}(\mathbf{w}) \quad \text{with} \quad \mathbf{w} = \epsilon^{-1}(\mathbf{x} - \mathbf{x}_0), \tag{2.62}$$

where $\mathbf{x}_0$ is an interior point of κ, ϵ is a positive constant, and $\boldsymbol{\xi}$ is compactly supported in a region D, the image of a strictly interior neighborhood $\kappa' \subset \kappa$ of $\mathbf{x}_0$ under the map $\mathbf{w}(\cdot)$. Accordingly, **u** vanishes on all of $\partial\kappa$, and on $\partial\kappa_y$ in particular, and is therefore admissible. For this choice (2.60) reduces, after dividing by ϵ^3 and passing to the limit $\epsilon \to 0$, to

$$\int_D \nabla\boldsymbol{\xi} \cdot \mathcal{A}[\nabla\boldsymbol{\xi}]dV \geq 0, \tag{2.63}$$

where $\mathcal{A} = \mathcal{M}_{|\mathbf{x}_0} = \mathcal{A}^t$. Here and henceforth ∇ is the gradient with respect to **w**.

We extend $\boldsymbol{\xi}$ to complex-valued vector fields as

$$\boldsymbol{\xi} = \boldsymbol{\xi}_1 + i\boldsymbol{\xi}_2, \tag{2.64}$$

where $\xi_{1,2}(\mathbf{w})$ are real-valued, and use these to derive

$$\nabla\xi \cdot \mathcal{A}[\nabla\bar{\xi}] = \nabla\xi_1 \cdot \mathcal{A}[\nabla\xi_1] + \nabla\xi_2 \cdot \mathcal{A}[\nabla\xi_2] + i(\ldots), \tag{2.65}$$

in which an overbar denotes complex conjugate.

Problem 2.7 Fill in the steps leading from (2.62) to (2.63). Show that the imaginary part $(\ldots)$ in (2.65) vanishes.

Thus, if (2.63) holds for real-valued ξ, then

$$\int_D \nabla\xi \cdot \mathcal{A}[\nabla\bar{\xi}]dV \geq 0 \tag{2.66}$$

for complex-valued ξ.

Consider

$$\xi(\mathbf{w}) = \mathbf{a}\exp(ik\mathbf{n}\cdot\mathbf{w})f(\mathbf{w}), \tag{2.67}$$

where $\mathbf{a}$ and $\mathbf{n}$ are real fixed vectors, k is a non-zero real number, and f is a real-valued differentiable function compactly supported in D. This yields

$$\nabla\xi = \exp(ik\mathbf{n}\cdot\mathbf{w})(ikf\mathbf{a}\otimes\mathbf{n} + \mathbf{a}\otimes\nabla f). \tag{2.68}$$

Substitution into (2.66) and division by k^2 results in

$$0 \leq \mathbf{a}\otimes\mathbf{n}\cdot\mathcal{A}[\mathbf{a}\otimes\mathbf{n}]\int_D f^2 dV + k^{-2}\int_D \mathbf{a}\otimes\nabla f\cdot\mathcal{A}[\mathbf{a}\otimes\nabla f]dV. \tag{2.69}$$

Finally, in the limit $k\to\infty$ we obtain

$$\mathbf{a}\otimes\mathbf{n}\cdot\mathcal{A}[\mathbf{a}\otimes\mathbf{n}] \geq 0, \tag{2.70}$$

which is just (2.61) on account of the arbitrariness of $\mathbf{x}_0$. We have thus shown that (2.61) is necessary for (2.60). The converse is not true, however, except in special circumstances. Our proof of the Legendre–Hadamard inequality is patterned after that found in the book by Giaquinta and Hildebrandt, which contains a wealth of valuable information about the calculus of variations.

Thus, if $\chi(\mathbf{x})$ is an energy minimizer, then it necessarily satisfies (2.61), at every point $\mathbf{x}\in\kappa$. If there is any point where (2.61) is violated, for some $\mathbf{a}\otimes\mathbf{n}$, then $\chi(\mathbf{x})$ is not an energy minimizer and is therefore unstable. Accordingly, there is considerable interest among elasticians in establishing conditions on the function Ψ which are such as to ensure that (2.61) is satisfied. See the books by Antman, Ciarlet, and Steigmann for discussions of those that have been proposed and studied. For our purposes, however, simpler restrictions will prove to be more relevant.

Consider the relation (1.42) between the Piola and Piola–Kirchhoff stresses. For hyperelastic materials this is equivalent to (2.32). We evaluate this on the one-parameter family $\chi(\mathbf{x}; u)$ and differentiate, obtaining

$$\Psi_{\mathbf{FF}}[\dot{\mathbf{F}}] = \dot{\mathbf{F}}\mathbf{S} + \mathbf{F}U_{\mathbf{EE}}[\dot{\mathbf{E}}], \tag{2.71}$$

where $\dot{\mathbf{E}}$ is given by (2.30). Here $U_{\mathbf{EE}}$ is the fourth-order tensor representing the second derivative of U with respect to the strain $\mathbf{E}$. This possesses major symmetry, as in (2.56). However, because $\mathbf{E}$ is symmetric, it also possesses the additional *minor symmetries*

$$\mathbf{A} \cdot U_{\mathbf{EE}}[\mathbf{B}] = \mathbf{A}^t \cdot U_{\mathbf{EE}}[\mathbf{B}] = \mathbf{A} \cdot U_{\mathbf{EE}}[\mathbf{B}^t]. \tag{2.72}$$

Accordingly,

$$U_{\mathbf{EE}}[\dot{\mathbf{E}}] = U_{\mathbf{EE}}[Sym(\mathbf{F}^t\dot{\mathbf{F}})] = U_{\mathbf{EE}}[\mathbf{F}^t\dot{\mathbf{F}}]; \tag{2.73}$$

that is, the skew part of $\mathbf{F}^t\dot{\mathbf{F}}$ is annihilated when operated on by $U_{\mathbf{EE}}$. Because $\dot{\mathbf{F}}$ is arbitrary, we may then use (2.71), with (2.55), to conclude that

$$\mathcal{M}[\mathbf{A}] = \mathbf{AS} + \mathbf{F}U_{\mathbf{EE}}[\mathbf{F}^t\mathbf{A}], \quad \text{for all} \quad \mathbf{A}. \tag{2.74}$$

Then,

$$\mathcal{M}[\mathbf{a} \otimes \mathbf{n}] = \mathbf{a} \otimes \mathbf{Sn} + \mathbf{F}U_{\mathbf{EE}}[\mathbf{F}^t\mathbf{a} \otimes \mathbf{n}], \tag{2.75}$$

and the Legendre–Hadamard inequality (2.61) reduces, with the aid of an identity proved in Problem 2.3, to

$$|\mathbf{a}|^2\, \mathbf{n} \cdot \mathbf{Sn} + \mathbf{F}^t\mathbf{a} \otimes \mathbf{n} \cdot U_{\mathbf{EE}}[\mathbf{F}^t\mathbf{a} \otimes \mathbf{n}] \geq 0. \tag{2.76}$$

Suppose the reference configuration itself to be both stable and stress free. We will see later that plasticity generally precludes the latter possibility, but it is nevertheless a basic premise of essentially all work on the subject of linear elasticity in which plasticity plays no role. In this case the classical *elastic modulus* tensor,

$$\mathcal{C}(\mathbf{x}) = U_{\mathbf{EE}|\mathbf{E}=\mathbf{0}}, \tag{2.77}$$

is such that

$$\mathbf{b} \otimes \mathbf{n} \cdot \mathcal{C}[\mathbf{b} \otimes \mathbf{n}] \geq 0, \tag{2.78}$$

where $\mathbf{a} = \mathbf{1b}$ in which $\mathbf{1}$ is the so-called *shifter*. This is like the identity, but not quite. Instead of mapping vectors in T_κ or T_{κ_t} to themselves, it simply converts vectors in the former into vectors in the latter. That is, it converts vectors belonging to T_κ into vectors belonging to T_{κ_t}, without otherwise altering them. Accordingly, $\mathbf{F}(\mathbf{x}, t) = \mathbf{1}$ when the body is undeformed, i.e., when κ_t momentarily coincides with κ. Because $\mathbf{C} = \mathbf{I}$ in this case, it follows that $\mathbf{1}^t\mathbf{1} = \mathbf{I}$, and hence that $\mathbf{1}$ is orthogonal. In fact, it is a rotation, since it is a possible value of $\mathbf{F}$; accordingly, it has a positive determinant, assuming, as we do, that κ is occupiable. Moreover, because $\mathbf{F}$ is the gradient of a field defined over the body, the rotation $\mathbf{1}$ is necessarily uniformly distributed and thus independent of $\mathbf{x}$. See the elegant proof given in Gurtin's book.

The notion of shifter was once more widely embraced but has since fallen out of favor. This has had the unfortunate and unintended consequence that fundamental deficiencies have been mistakenly attributed to linear elasticity theory, most notably the allegation that it fails to satisfy the requirement of frame invariance. This incorrect view can be traced to a misinterpretation of the shifter as a conventional identity. One wonders how teachers of the subject could sleep at night, with the unquestioned success of linear elasticity as a predictive theory on the one hand, and its supposed failure to be frame invariant—and thus wrong—on the other. Unfortunately, typical pedagogical practice is to sweep this puzzle under the rug by simply omitting any mention of it.

We shall have occasion to make use of the shifter concept in due course. In the meantime, the interested reader may find excellent treatments of it in the article by Ericksen, published as an appendix to the authoritative treatise by Truesdell and Toupin, and in the book by Marsden and Hughes.

Returning to the matter at hand, it is conventional, and also realistic for essentially all solids of interest, to ensure that (2.78) is satisfied by requiring $\mathcal{C}$ to be such that

$$\mathbf{A} \cdot \mathcal{C}[\mathbf{A}] > 0 \quad \text{for all non-zero symmetric} \quad \mathbf{A} \tag{2.79}$$

that map T_κ to itself. This implies that $\mathbf{A} \cdot \mathcal{C}[\mathbf{A}] \geq 0$ for all $\mathbf{A}$, symmetric or not (why?), and this in turn guarantees that (2.60) is satisfied when the body is undeformed, i.e.,

$$\int_\kappa \nabla\mathbf{u} \cdot \mathcal{M}(\mathbf{1}; \mathbf{x})[\nabla\mathbf{u}]dV = \int_\kappa \mathbf{1}^t\nabla\mathbf{u} \cdot \mathcal{C}(\mathbf{x})[\mathbf{1}^t\nabla\mathbf{u}]dV \geq 0. \tag{2.80}$$

If $|\mathbf{E}|$ is small, and if the stress vanishes when $\mathbf{E}$ vanishes, then the Piola–Kirchhoff stress is given to leading order by the classical linear stress–strain relation

$$\mathbf{S} = \mathcal{C}[\mathbf{E}]. \tag{2.81}$$

Inequality (2.79) is the statement that $\mathcal{C}$ is positive definite, and hence invertible, on the linear space of symmetric tensors. Then (2.81) is invertible, i.e.,

$$\mathbf{E} = \mathcal{L}[\mathbf{S}], \tag{2.82}$$

where $\mathcal{L} = \mathcal{C}^{-1}$, not to be confused with the load potential, is the fourth-order *elastic compliance* tensor. This possesses major symmetry and the two minor symmetries and is also positive definite on the space of symmetric tensors.

Problem 2.8 Prove these statements.

2.4 Material symmetry

The concept of material symmetry plays a crucial role in elasticity theory and, as we shall see later, in plasticity theory too. This has to do with the idea that there may be certain non-trivial transformations of a neighborhood $N_{\kappa}(p)$ of a material point p that are undetectable by experiment. These are called *material symmetry* transformations. They are of fundamental importance to an understanding of material response.

To set the stage for this discussion, consider a change of reference configuration. Instead of using κ as reference, let us use the fixed configuration μ, say. As usual the position of a material point p in κ is denoted by $\mathbf{x}$; let $\mathbf{u}$ be its position in μ. We suppose, as usual, that both configurations stand in one-to-one correspondence with the body B, and hence that they are in such correspondence with each other as well. Thus, there exists an invertible map $\boldsymbol{\lambda}$, say, such that

$$\mathbf{u} = \boldsymbol{\lambda}(\mathbf{x}), \tag{2.83}$$

which, if smooth, has an invertible gradient

$$\mathbf{R} = \nabla\boldsymbol{\lambda}. \tag{2.84}$$

To avoid ambiguity we must reinstate subscripts on constitutive functions that involve the deformation gradient, as these are computed on the basis of a specified reference configuration. Thus, for example, the constitutive function giving the value of the strain energy per unit volume of κ_t, based on κ as reference, is $\psi_{\kappa}(\mathbf{F}_{\kappa}; \mathbf{x})$, while that giving the *same* value, but based on μ, is $\psi_{\mu}(\mathbf{F}_{\mu}; \mathbf{u})$. The actual configuration κ_t has not been affected. Thus,

$$\mathbf{F}_{\kappa} d\mathbf{x} = d\mathbf{y} = \mathbf{F}_{\mu} d\mathbf{u} = \mathbf{F}_{\mu}\mathbf{R} d\mathbf{x}, \tag{2.85}$$

and therefore

$$\mathbf{F}_{\kappa} = \mathbf{F}_{\mu}\mathbf{R}. \tag{2.86}$$

The energy content has not been affected either. Accordingly,

$$\psi_{\kappa}(\mathbf{F}_{\kappa}; \mathbf{x}) = \psi_{\mu}(\mathbf{F}_{\mu}; \mathbf{u}), \tag{2.87}$$

where

$$\psi_{\mu}(\mathbf{F}_{\mu}; \mathbf{u}) = \psi_{\kappa}(\mathbf{F}_{\mu}\mathbf{R}(\boldsymbol{\lambda}^{-1}(\mathbf{u})); \boldsymbol{\lambda}^{-1}(\mathbf{u})). \tag{2.88}$$

This indicates how the constitutive *function* changes when we choose a different reference configuration. Note that the *definition* of a hyperelastic material, while involving a reference configuration, is independent of the particular reference configuration adopted.

Having deduced the formula for the change of constitutive function attending a change of reference, consider next a *local* change of reference in which an open neighborhood $N_{\kappa}(p)$ is transformed into another open neighborhood $N_{\mu}(p)$ without affecting the remainder of the body. We can think of surgically removing $N_{\kappa}(p)$, transforming it to $N_{\mu}(p)$, and reinserting it in the body. Taking these neighborhoods to be open means we don't need to worry about what happens at their boundaries in this thought experiment, as there are none. Again let $\boldsymbol{\lambda}$ be the smooth invertible map from the former to the latter; that is, the position $\mathbf{u}$ of a material point in $N_{\mu}(p)$ is given by (2.83), where $\mathbf{x}$ is the position of the same point in $N_{\kappa}(p)$. We arrange $\boldsymbol{\lambda}$ such that it preserves the location of p, so that if $\mathbf{x}_0$ is the position of p in $N_{\kappa}(p)$, then it is also its position in $N_{\mu}(p)$, i.e., $\mathbf{x}_0 = \boldsymbol{\lambda}(\mathbf{x}_0)$. Then (2.87), evaluated at p, yields

$$\psi_{\kappa}(\mathbf{F}_{\kappa}; \mathbf{x}_0) = \psi_{\mu}(\mathbf{F}_{\mu}; \mathbf{x}_0), \quad \text{where} \quad \mathbf{F}_{\kappa} = \mathbf{F}_{\mu}\mathbf{R}_0 \quad \text{with} \quad \mathbf{R}_0 = \nabla\boldsymbol{\lambda}_{|\mathbf{x}_0}. \tag{2.89}$$

We say that $\boldsymbol{\lambda}$ generates a *symmetry transformation at* the point p if it is such that the mechanical responses of $N_{\mu}(p)$ and $N_{\kappa}(p)$ to a given deformation coincide *at* p; that is, if

$$\psi_{\kappa}(\mathbf{F}; \mathbf{x}_0) = \psi_{\mu}(\mathbf{F}; \mathbf{x}_0) \tag{2.90}$$

for any possible deformation gradient $\mathbf{F}$. This means that if $N_{\kappa}(p)$ is occupiable, and hence if $J_F > 0$, then we should arrange $\boldsymbol{\lambda}$ such that $N_{\mu}(p)$ is too; that is, such that $J_R > 0$. The italics are used here to emphasize the fact that this discussion pertains to a *single* point p. Combining (2.89) and (2.90), we arrive at the restriction

$$\psi_{\kappa}(\mathbf{F}; \mathbf{x}) = \psi_{\kappa}(\mathbf{F}\mathbf{R}; \mathbf{x}) \tag{2.91}$$

on the single function ψ_{κ}, in which the subscript $_0$ has been dropped to emphasize the fact that p, though fixed, is an arbitrary point of the body.

Let $g_{\kappa(p)}$ be the set of tensors such that (2.91) is satisfied:

$$g_{\kappa(p)} = \{\mathbf{R}(\mathbf{x}) \mid \psi_{\kappa}(\mathbf{F}; \mathbf{x}) = \psi_{\kappa}(\mathbf{F}\mathbf{R}(\mathbf{x}); \mathbf{x})\}. \tag{2.92}$$

This set again pertains to the particular point p. Accordingly, the notion of material symmetry is local in the sense that the set of $\mathbf{R}$'s satisfying (2.91), for a given $\mathbf{F}$, generally varies from one material point to another. For this reason it is necessary to identify the material point p to which this set pertains.

Before proceeding we address a subtle point that often leads to confusion. In (2.92) $\mathbf{R}(\mathbf{x})$ need not be a gradient *field*; that is, it should not be interpreted as the gradient of a single map defined over the entire body. This is because the map $\boldsymbol{\lambda}$ used in the definition of material symmetry has been defined on a neighborhood of p. The maps themselves may vary from one neighborhood to another and hence from one material point to another. In this way we allow symmetry groups at different material points to be unrelated, and thus accommodate non-uniform composites, for example, in which the materials constituting a body may have entirely different properties. As we shall see, this issue has major consequences for plasticity theory in particular.

It is easy to show that the set defined by (2.92) is a *group* (see Steigmann, for example). This means, among other things, that the product of two members of $g_{\mathcal{K}(p)}$ is also a member. It then follows that $\mathbf{R}^n$ is a member for any positive integer n, i.e., $\psi_{\mathcal{K}}(\mathbf{F};\mathbf{x}) = \psi_{\mathcal{K}}(\mathbf{F}\mathbf{R}^n;\mathbf{x})$. Noting that $\det(\mathbf{F}\mathbf{R}^n) = J_F \det(\mathbf{R}^n) = J_F J_R^n$, we conclude that the energy remains unaffected by unbounded dilation ($J_R > 1$ and $n \to \infty$) or compaction ($J_R < 1$ and $n \to \infty$). This plainly nonsensical conclusion can only be avoided by imposing $J_R = 1$. We thus require that $g_{\mathcal{K}(p)}$ be contained in the *unimodular* group u, i.e.,

$$g_{\mathcal{K}(p)} \subset u = \{\mathbf{R} \mid J_R = 1\}. \tag{2.93}$$

Following Noll, we define our elastic material point p to be a *solid point* if there exists a *local* configuration $\kappa^*(p)$; that is, a neighborhood of p, such that

$$g_{\mathcal{K}^*(p)} \subset Orth^+, \tag{2.94}$$

the group of proper-orthogonal tensors, or rotations, defined by

$$Orth^+ = \{\mathbf{R} \in u \mid \mathbf{R}^t\mathbf{R} = \mathbf{I}\}. \tag{2.95}$$

Note that $\kappa^*(p)$ need have no relation to the reference configuration κ used for the purposes of analysis. Rather, the definition is motivated by the fact that, in the case of crystalline solids, for example, there exists an *undistorted* configuration of the crystal which is such that a unit cell of the lattice is mapped to itself by certain rotations characterizing its symmetry. In the general case the same definition may be motivated by the observation that a change of shape, i.e., a strain induced by the map $\boldsymbol{\lambda}$, is mechanically detectable in solids, whereas symmetry transformations should be undetectable. Accordingly, the Cauchy–Green deformation tensor induced by $\boldsymbol{\lambda}$, namely $\mathbf{R}^t\mathbf{R}$, should be fixed at the value $\mathbf{I}$. *Isotropic* solids are those for which there exists $\kappa^*(p)$ such that $g_{\mathcal{K}^*(p)} = Orth^+$, meaning that the constitutive function relative to $\kappa^*(p)$ is insensitive to *any* rotation preceding the application of the deformation.

It stands to reason that the symmetry group relative to a distorted configuration does not generally belong to $Orth^+$. A simple illustration is afforded by the unit cell of a cubic crystal. The cell is in the shape of a cube when the lattice is undistorted, and is then mapped to itself by certain discrete rotations that do not affect its mechanical response. This is no longer the case if the lattice is first sheared, for example. Rather, the symmetry group relative to the sheared configuration is given by Noll's rule.

To derive this rule consider two references κ_1 and κ_2, and let $\boldsymbol{\pi}$ be a smooth invertible map, with gradient $\mathbf{K}$, from the first to the second. We regard these references as local neighborhoods of the material point p in two fixed global configurations. Let p have position $\mathbf{x}_1$ in κ_1 and $\mathbf{x}_2$ in κ_2, so that $\mathbf{x}_2 = \boldsymbol{\pi}(\mathbf{x}_1)$. Then from the change-of-reference formula (2.87) we have $\psi_{\kappa_2}(\mathbf{F}; \mathbf{x}_2) = \psi_{\kappa_1}(\mathbf{FK}(\mathbf{x}_1); \mathbf{x}_1)$; equivalently,

$$\psi_{\kappa_1}(\mathbf{F}; \boldsymbol{\pi}^{-1}(\mathbf{x}_2)) = \psi_{\kappa_2}(\mathbf{FK}^{-1}; \mathbf{x}_2). \tag{2.96}$$

Suppose that $\mathbf{R} \in g_{\kappa_1(p)}$, i.e.,

$$\psi_{\kappa_1}(\mathbf{F}; \boldsymbol{\pi}^{-1}(\mathbf{x}_2)) = \psi_{\kappa_1}(\mathbf{FR}; \boldsymbol{\pi}^{-1}(\mathbf{x}_2)), \tag{2.97}$$

where, from (2.96), $\psi_{\kappa_1}(\mathbf{FR}; \boldsymbol{\pi}^{-1}(\mathbf{x}_2)) = \psi_{\kappa_2}(\mathbf{FRK}^{-1}; \mathbf{x}_2)$. Then (2.96) and (2.97) combine to yield

$$\psi_{\kappa_2}(\mathbf{FRK}^{-1}; \mathbf{x}_2) = \psi_{\kappa_1}(\mathbf{F}; \boldsymbol{\pi}^{-1}(\mathbf{x}_2)) = \psi_{\kappa_2}(\mathbf{FK}^{-1}; \mathbf{x}_2), \tag{2.98}$$

which is the same as

$$\psi_{\kappa_2}(\hat{\mathbf{F}}; \mathbf{x}_2) = \psi_{\kappa_2}(\hat{\mathbf{F}}\mathbf{KRK}^{-1}; \mathbf{x}_2), \quad \text{where} \quad \hat{\mathbf{F}} = \mathbf{FK}^{-1}. \tag{2.99}$$

We thus arrive at Noll's rule:

$$g_{\kappa_2(p)} = \{\mathbf{KRK}^{-1} \mid \mathbf{R} \in g_{\kappa_1(p)}\}. \tag{2.100}$$

Note once again that this discussion pertains to a single point p. This means that $\mathbf{K}$ is not a gradient *field*, but is instead specific to p. Again this is because the map $\boldsymbol{\pi}$ underpinning this discussion is defined on a neighborhood of p, and the maps themselves may thus vary from one material point to another, each with its own neighborhood. In other words, (2.100) is meaningful *whether or not* a single global map $\boldsymbol{\pi}$ exists for the entire body such that $\mathbf{K} = \nabla\boldsymbol{\pi}_{|\mathbf{x}_1}$. Note also that $g_{\kappa_2(p)} \subset u$, but if $g_{\kappa_1(p)}$ is contained in $Orth^+$ then $g_{\kappa_2(p)}$ is not, unless $\mathbf{K}$ is suitably restricted.

Problem 2.9 Suppose an elastic material is isotropic, at a particular material point p, relative to reference configuration κ_1. Thus, $g_{\kappa_1(p)} = Orth^+$. Consider a map from configuration κ_1 to another configuration κ_2 given by $\boldsymbol{\pi}(\mathbf{x}) = \lambda\mathbf{k}(\mathbf{k}\cdot\mathbf{x}) + \lambda^{-1/2}[\mathbf{x} - \mathbf{k}(\mathbf{k}\cdot\mathbf{x})]$ for all $\mathbf{x} \in \kappa_1$, where λ is a positive constant and $\mathbf{k}$ is a constant unit vector.

(a) Consider a rotation $\mathbf{R} \in g_{\kappa_1(p)}$. What is the corresponding element of $g_{\kappa_2(p)}$? Is the material isotropic relative to configuration κ_2?

(b) Suppose the material is transversely isotropic relative to κ_1, i.e., $g_{\kappa_1(p)} = \{\mathbf{R} \in Orth^+ \text{ and } \mathbf{Re} = \mathbf{e}\}$ for some unit vector $\mathbf{e}$. The response of the material is then insensitive to all rotations about the axis $\mathbf{e}$. Show that the material is also transversely isotropic relative to κ_2 if $\mathbf{e} = \mathbf{k}$. (Hint: Use the Rodrigues representation formula for rotations, discussed, for example, in the book by Chadwick.)

References

Antman, S. S. (2005). *Nonlinear Problems of Elasticity.* Springer, Berlin.

Chadwick, P. (1976). *Continuum Mechanics: Concise Theory and Problems.* Dover, New York.

Ciarlet, P .G. (1988). *Mathematical Elasticity*, Vol 1: *Three Dimensional Elasticity.* North-Holland, Amsterdam.

Ericksen, J. L. (1960). Tensor fields, appendix to: Truesdell, C., and Toupin, R. A., The classical field theories, in Flügge, S. (Ed.), *Handbuch der Physik*, Vol. III/1, pp. 794–858. Springer, Berlin.

Fleming, W. (1977). *Functions of Several Variables.* Springer, Berlin.

Giaquinta, M., and Hildebrandt, S. (2004). *Calculus of Variations I.* Springer, Berlin.

Gurtin, M. E. (1981). *An Introduction to Continuum Mechanics.* Academic Press, Orlando.

Marsden, J. E., and Hughes, T. J. R. (1994). *Mathematical Foundations of Elasticity.* Dover, New York.

Murdoch. A. I. (2003). Objectivity in classical continuum physics: A rationale for discarding the "principle of invariance under superposed rigid-body motions" in favor of purely objective considerations. *Continuum Mech. Thermodyn.* 15, 309–20.

Noll, W. (1958). A mathematical theory of the mechanical behavior of continuous media. *Arch. Ration. Mech. Anal.* 2, 197–226.

Steigmann, D. J. (2017). *Finite Elasticity Theory.* Oxford University Press, Oxford.

3

A primer on tensor analysis in three-dimensional space

Certain concepts that we will encounter in our study of the kinematics of elastic–plastic deformation, such as the notion of a dislocation density, may be framed in terms of differential geometry, a proper understanding of which requires a facility with tensor analysis. The geometric interpretations we will encounter are profound and beautiful, and promote an understanding of our subject from a perspective that unifies it with other branches of the physical sciences. To prepare for this, we devote the present chapter to a survey of tensor analysis and differential geometry in three-dimensional space. We emphasize the Euclidean space of our common experience; on the one hand, to develop intuition, and on the other, because this affords a relatively easy segue to concepts from non-Euclidean geometry that are of immediate relevance to the phenomena with which we are concerned.

Most of this chapter is normally part and parcel of the background acquired by the beginning graduate student. Indeed the literature on this subject is vast, and shows no sign of abating. Accordingly, we will be brief. Thorough discussions of any concept encountered here that may be unfamiliar may be found in the excellent texts by Sokolnikoff, Lichnerowicz, Green and Zerna, Flügge, and Simmonds, among many others. A lucid account of advanced topics pertaining to *affine geometry*, including those geometries of relevance here, may be found in the books by Schouten, Lovelock and Rund, and Szerkeres, the last of these coming from a more modern perspective.

3.1 Coordinates, bases, vectors, and metrics

As a practical matter, we may think of Euclidean 3-space as consisting of positions $\mathbf{y}$, say, measured from a specified origin, that locate the points of the space. Each point is identified by a triplet $\{\xi^i\}; i = 1, 2, 3$. The elements of the triplet are the *coordinates* of the point, and the position, varying as it does from one point to another, is accordingly a function of the coordinates; we write $\mathbf{y}(\xi^i)$, the intended meaning being that $\mathbf{y}$ is a function of all three coordinates. We shall assume this position field to be as smooth as required by the ensuing development.

A Course on Plasticity Theory. David J. Steigmann, Oxford University Press. © David J. Steigmann (2022).
DOI: 10.1093/oso/9780192883155.003.0003

Let $\{c^i\}$ be a collection of three constants. Then, for example, $\mathbf{y}(\xi^1, c^2, c^3)$ is the parametric representation of a particular ξ^1 - coordinate curve. The derivative

$$\mathbf{g}_1 = \partial\mathbf{y}/\partial\xi^1 = \lim_{\Delta\xi\to 0}\frac{1}{\Delta\xi}[\mathbf{y}(c^1+\Delta\xi, c^2, c^3) - \mathbf{y}(c^1, c^2, c^3)] \tag{3.1}$$

is the limit of the scaled chord connecting the points on this curve having coordinates c^1 and $c^1 + \Delta\xi$ and is therefore tangential to the curve at the point with coordinate c^1. If $\mathbf{g}_1$ is a continuous function of ξ^1, then the curve is smooth. You can convince yourself that $\mathbf{g}_1$ is oriented in the direction of increasing ξ^1. In the same way we define the three vector fields

$$\mathbf{g}_i = \mathbf{y}_{,i} \equiv \partial\mathbf{y}/\partial\xi^i, \tag{3.2}$$

in which the comma notation affords a convenient way to denote the partial derivative with respect to the ith coordinate. The three vectors $\mathbf{g}_1, \mathbf{g}_2$, and $\mathbf{g}_3$ are then tangential to the three coordinate curves $\mathbf{y}(\xi^1, c^2, c^3)$, $\mathbf{y}(c^1, \xi^2, c^3)$, and $\mathbf{y}(c^1, c^2, \xi^3)$, respectively, at the point with coordinates $\xi^i = c^i$.

Henceforth we assume the three coordinates to be independent in the sense that the induced tangent vectors $\mathbf{g}_i$ are linearly independent. This is the case if and only if the scalar triple product $\mathbf{g}_1 \cdot \mathbf{g}_2 \times \mathbf{g}_3$ is non-zero. For the sake of convenience we restrict ourselves to right-handed coordinates for which $\mathbf{g}_1 \cdot \mathbf{g}_2 \times \mathbf{g}_3 > 0$. In general, for a given coordinate system, linear independence cannot be guaranteed to hold globally. There may be particular points, curves, or surfaces where this condition breaks down. Examples are furnished by the origin in a system of plane polar coordinates, or the axis of a system of cylindrical polar coordinates. These singularities occur where the relation between the position $\mathbf{y}$ and the coordinates ξ^i ceases to be one-to-one. To characterize them, we introduce global right-handed Cartesian coordinates y^i, in terms of which position is represented simply as $\mathbf{y} = y^k\mathbf{i}_k$, where the $\mathbf{i}_j$ are constant orthonormal unit vectors aligned with the three Cartesian coordinate lines, and a sum over diagonally repeated indices from 1 to 3 is implied. This is Einstein's famous *summation convention*, invented to dramatically declutter formulae encountered in relativity theory. Because the coordinates y^k are uniquely determined by $\mathbf{y}(\xi^i)$, they are also functions $\hat{y}^k$ of the ξ^i, and we have the useful result

$$\mathbf{g}_i = y^k_{,i}\mathbf{i}_k, \quad \text{where} \quad y^k_{,i} = \partial y^k/\partial\xi^i, \tag{3.3}$$

in which the superposed caret is suppressed to avoid clutter, yielding the $\mathbf{g}_i$ when the functions $y^k = \hat{y}^k(\xi^i)$ are known. Again we have a sum on k for each fixed i. Introducing the permutation symbol

$$e_{ijk} = \mathbf{i}_i \cdot \mathbf{i}_j \times \mathbf{i}_k = \left\{\begin{array}{c} 1, \text{if } \{ijk\} \text{ is an even permutation of } \{123\} \\ -1, \text{if } \{ijk\} \text{ is an odd permutation of } \{123\} \\ 0, \text{otherwise} \end{array}\right\}, \tag{3.4}$$

we then have

$$\mathbf{g}_1 \cdot \mathbf{g}_2 \times \mathbf{g}_3 = y^i_{,1}\mathbf{i}_i \cdot y^j_{,2}\mathbf{i}_j \times y^k_{,3}\mathbf{i}_k = e_{ijk}y^i_{,1}y^j_{,2}y^k_{,3} = \det(y^i_{,j}), \tag{3.5}$$

where $(y^i_{,j})$ is the 3×3 matrix with (i, j) entry equal to $y^i_{,j}$. The inverse function theorem guarantees that the relation between the y's and the ξ's, and hence that between the latter and $\mathbf{y}$, is locally one-to-one wherever this determinant is non-zero. Accordingly, the singularities just mentioned occur wherever this determinant vanishes. At regular points where the determinant is non-zero, we can thus invert the function $\mathbf{y}(\xi^i)$ to obtain $\xi^i = \hat{\xi}^i(\mathbf{y})$. Moreover, the coordinates ξ^i are right-handed if and only if

$$\det(y^i_{,j}) > 0, \tag{3.6}$$

a restriction that we assume henceforth to hold almost everywhere in 3-space, i.e., on sets of positive volume measure.

At this stage we recall that the gradient of a scalar-valued function $f(\mathbf{y})$ is the unique vector *gradf* such that

$$df = \mathit{grad}f \cdot d\mathbf{y}, \tag{3.7}$$

which is shorthand for the formula

$$f' = \mathit{grad}f \cdot \mathbf{y}', \tag{3.8}$$

wherein

$$\mathbf{y}' = \frac{d}{dt}\hat{\mathbf{y}}(t), \quad f' = \frac{d}{dt}f(\hat{\mathbf{y}}(t)) \tag{3.9}$$

and $\hat{\mathbf{y}}(t) = \mathbf{y}(\xi^i(t))$ is the parametric representation of an arbitrary curve. By the chain rule and (3.7), applied to the coordinate functions,

$$df = f_{,i}d\xi^i = f_{,i}\mathit{grad}\hat{\xi}^i \cdot d\mathbf{y}, \tag{3.10}$$

where, as usual, $f_{,i} = \partial f/\partial \xi^i$. Substituting this in place of the left-hand side of (3.7) and invoking the arbitrariness of the curve, i.e., the arbitrariness of $d\mathbf{y}$, we find that

$$\mathit{grad}f = f_{,i}\mathbf{g}^i, \quad \text{where} \quad \mathbf{g}^i = \mathit{grad}\xi^i, \tag{3.11}$$

in which the superposed caret has again been dropped. Evidently $\mathbf{g}^1$ is orthogonal to the surface defined by $\xi^1 = c^1$, etc.

Cartesian coordinates have the interesting property that

$$\mathbf{i}_k = \partial \mathbf{y}/\partial y^k \quad \text{and} \quad \mathbf{i}_k = grad y^k. \tag{3.12}$$

Regarding the ξ^i as functions of the y^k, we can use this fact together with the chain rule to obtain

$$\mathbf{g}^i = \frac{\partial \xi^i}{\partial y^k} grad y^k = \frac{\partial \xi^i}{\partial y^k}\mathbf{i}_k, \tag{3.13}$$

which may be compared to (3.3) and in which a sum on k is again implied. Here and henceforth we write out the partial derivative explicitly when differentiating with respect to Cartesian coordinates. From this we can conclude that the set $\{\mathbf{g}^i\}$ is linearly independent wherever $\{\mathbf{g}_i\}$ is, and with the same orientation. Thus,

$$\mathbf{g}^1 \cdot \mathbf{g}^2 \times \mathbf{g}^3 = \frac{\partial \xi^1}{\partial y^i}\mathbf{i}_i \cdot \frac{\partial \xi^2}{\partial y^j}\mathbf{i}_j \times \frac{\partial \xi^3}{\partial y^k}\mathbf{i}_k = \det(\partial \xi^i/y^j) = [\det(y^i_{,j})]^{-1}, \tag{3.14}$$

which is non-zero if and only if $\mathbf{g}_1 \cdot \mathbf{g}_2 \times \mathbf{g}_3$ is non-zero, and with the same sign. Accordingly, $\{\mathbf{g}^i\}$ constitutes a basis for 3-vectors wherever $\{\mathbf{g}_i\}$ does, i.e., wherever $\det(y^i_{,j}) \neq 0$. Invoking (3.3), (3.13), and the chain rule, we also find that

$$\mathbf{g}^i \cdot \mathbf{g}_j = \frac{\partial \xi^i}{\partial y^k}\mathbf{i}_k \cdot y^l_{,j}\mathbf{i}_l = \frac{\partial \xi^i}{\partial y^k}y^k_{,j} = \xi^i_{,j} = \delta^i_j \equiv \begin{Bmatrix} 1, \text{if } i = j \\ 0, \text{otherwise} \end{Bmatrix}. \tag{3.15}$$

Here δ^i_j, the components of the Kronecker delta, are the elements of the 3×3 unit matrix, and we have used the fact that $\mathbf{i}_k \cdot \mathbf{i}_l$ equals unity if $k = l$, and zero otherwise. Because of this result the basis $\{\mathbf{g}^i\}$ is said to be *reciprocal* to $\{\mathbf{g}_i\}$, and vice versa. It is clear from (3.12) that the induced tangent vectors coincide with their reciprocals when the coordinates are Cartesian.

We will have occasion to make use of tensorial generalizations of the permutation symbol. We define these, using Cartesian coordinates as intermediaries, by

$$\epsilon_{rst} = y^i_{,r}y^j_{,s}y^k_{,t}e_{ijk} \quad \text{and} \quad \epsilon^{rst} = \frac{\partial \xi^r}{\partial y^i}\frac{\partial \xi^s}{\partial y^j}\frac{\partial \xi^t}{\partial y^k}e^{ijk}, \quad \text{where} \quad e^{ijk} = e_{ijk}. \tag{3.16}$$

Thus,

$$\epsilon_{rst} = \det(y^i_{,j})e_{rst} \quad \text{and} \quad \epsilon^{rst} = [\det(y^i_{,j})]^{-1}e^{rst}. \tag{3.17}$$

These enter in a natural way in cross-product operations. For example, using (3.3) and the chain rule gives

$$\begin{aligned} \mathbf{g}_i \times \mathbf{g}_j &= y^k_{,i}y^l_{,j}\mathbf{i}_k \times \mathbf{i}_l = y^k_{,i}y^l_{,j}e_{klm}\mathbf{i}_m \\ &= y^k_{,i}y^l_{,j}e_{klm}\, grad y^m = y^k_{,i}y^l_{,j}y^m_{,r}e_{klm}\, grad \xi^r, \end{aligned} \tag{3.18}$$

which is just

$$\mathbf{g}_i \times \mathbf{g}_j = \epsilon_{ijr}\mathbf{g}^r. \tag{3.19}$$

In the same way we can establish that

$$\mathbf{g}^i \times \mathbf{g}^j = \epsilon^{ijr}\mathbf{g}_r. \tag{3.20}$$

These are the extensions of the Cartesian formula $\mathbf{i}_i \times \mathbf{i}_j = e_{ijr}\mathbf{i}_r$ to arbitrary right-handed coordinate systems.

Our next order of business is to define the *metric* components

$$g_{ij} = \mathbf{g}_i \cdot \mathbf{g}_j. \tag{3.21}$$

Using the reciprocal basis, we may also define reciprocal metric components

$$g^{ij} = \mathbf{g}^i \cdot \mathbf{g}^j. \tag{3.22}$$

These obviously possess the symmetries $g_{ij} = g_{ji}$ and $g^{ij} = g^{ji}$. Further, we can exploit the fact that $\{\mathbf{g}^j\}$ is a basis for 3-vectors under our hypotheses to write $\mathbf{g}_i$ in the form $\mathbf{g}_i = a_{ij}\mathbf{g}^j$ with a unique set $\{a_{ij}\}$ of coefficients. Then,

$$a_{ik} = a_{ij}\delta^j_k = a_{ij}\mathbf{g}^j \cdot \mathbf{g}_k = \mathbf{g}_i \cdot \mathbf{g}_k = g_{ik}, \tag{3.23}$$

implying that

$$\mathbf{g}_i = g_{ij}\mathbf{g}^j. \tag{3.24}$$

In the same way we have

$$\mathbf{g}^i = g^{ij}\mathbf{g}_j. \tag{3.25}$$

We may combine these results as

$$0 = \delta^i_k\mathbf{g}^k - \mathbf{g}^i = \delta^i_k\mathbf{g}^k - g^{ij}\mathbf{g}_j = (\delta^i_k - g^{ij}g_{jk})\mathbf{g}^k, \tag{3.26}$$

and conclude, by the presumed linear independence of the $\mathbf{g}^i$, that

$$g^{ij}g_{jk} = \delta^i_k, \tag{3.27}$$

and hence that the matrices (g^{ij}) and (g_{ij}) are mutual inverses.

This observation begs an obvious question concerning the invertibility of these matrices. To address this, let

$$g = \det(g_{ij}). \tag{3.28}$$

From (3.3) and (3.21) we have

$$g_{ij} = y^k_{,i} y^l_{,j} \mathbf{i}_k \cdot \mathbf{i}_l = \sum_k y^k_{,i} y^k_{,j} \tag{3.29}$$

which implies that $g = [\det(y^i_{,j})]^2$ and hence, with (3.6), that (g_{ij}) is indeed invertible; and that

$$g^{-1} = \det(g^{ij}). \tag{3.30}$$

The name *metric* derives from $d\mathbf{y} = \mathbf{g}_i d\xi^i$ (see (3.2)) and $ds = |d\mathbf{y}|$, yielding

$$ds^2 = g_{ij} d\xi^i d\xi^j. \tag{3.31}$$

Integrating ds along a curve with parametric equations $\xi^i(t)$, from t_1 to t_2, thus yields the arclength of the part of the curve connecting the latter two points. Further, if $d\mathbf{y} \neq \mathbf{0}$ then $ds^2 > 0$, and by the linear independence of $\{\mathbf{g}_i\}$, at least one element of $\{d\xi^i\}$ is non-zero. Equation (3.31) then implies that (g_{ij}) is a positive definite matrix and hence that $g > 0$.

As a side note we conclude from (3.17) that

$$\epsilon_{ijk} = \sqrt{g} e_{ijk} \quad \text{and} \quad \epsilon^{ijk} = e^{ijk} / \sqrt{g} \tag{3.32}$$

in our right-handed coordinate system.

We may exploit the status of $\{\mathbf{g}_i\}$ and $\{\mathbf{g}^i\}$ as bases for 3-vectors to write any such vector, $\mathbf{v}$ say, in either of the forms

$$\mathbf{v} = v_i \mathbf{g}^i = v^i \mathbf{g}_i, \tag{3.33}$$

with unique, and generally distinct, coefficients v_i and v^i. These are classically referred to as the *co-* and *contravariant* components of $\mathbf{v}$, respectively, and we shall adhere to this convention. For example, using the second representation,

$$\mathbf{v} \cdot \mathbf{g}^j = v^i \mathbf{g}_i \cdot \mathbf{g}^j = v^i \delta_i^j = v^j, \tag{3.34}$$

so that contravariant components are simply the projections of the vector in question onto the corresponding reciprocal basis elements. The first representation similarly yields

$$\mathbf{v} \cdot \mathbf{g}_j = v_j \tag{3.35}$$

and hence the conclusion that the covariant components are its projections onto the tangent basis elements.

Beyond this we can use the two representations in (3.33) to infer that

$$v_j \mathbf{g}^j = v^i \mathbf{g}_i = v^i g_{ij} \mathbf{g}^j, \tag{3.36}$$

and hence that

$$v_j = g_{ji} v^i, \tag{3.37}$$

where we have used $g_{ji} = g_{ij}$. Alternatively, we can again proceed from (3.33), or directly from the invertibility of the matrix of metric components, to derive the companion formula

$$v^j = g^{ji} v_i. \tag{3.38}$$

These rules go by the names *lowering the index* and *raising the index*, respectively.

Problem 3.1 Compute the $\mathbf{g}_i$, g_{ij}, g^{ij}, and $\mathbf{g}^i$ in spherical coordinates $\{\xi^i\} = \{r, \theta, \phi\}$, using

$$y^1 = r\cos\phi\cos\theta, \quad y^2 = r\cos\phi\sin\theta, \quad y^3 = r\sin\phi.$$

At what points, if any, are the relations between the Cartesian and spherical coordinates non-invertible?

Problem 3.2 Establish the four representations

$$u^i v^j g_{ij} = u^i v_i = u_j v_i g^{ij} = u_j v^j$$

for the scalar product $\mathbf{u} \cdot \mathbf{v}$ of two 3-vectors.

Problem 3.3 Establish the two representations

$$\epsilon_{kij} u^i v^j \mathbf{g}^k = \epsilon^{kij} u_i v_j \mathbf{g}_k$$

for the cross product $\mathbf{u} \times \mathbf{v}$ of two 3-vectors.

Problem 3.4 Find $\{\mathbf{g}_i\}$, $\{\mathbf{g}^i\}$, g_{ij}, and g^{ij} in terms of $\{\xi^i\} = \{\xi, \eta, \phi\}$ for the following coordinate systems: (a) Parabolic coordinates: $y^1 = \xi\eta\cos\phi$, $y^2 = \xi\eta\sin\phi$, $y^3 = \frac{1}{2}(\xi^2 - \eta^2)$. (b) Elliptic-cylindrical coordinates: $y^1 = \cosh\xi\cos\eta$, $y^2 = \sinh\xi\sin\eta$, $y^3 = \phi$.

Problem 3.5 Consider the complex-valued analytic function $g(z)$ of the complex variable $z = x + iy$, where $x(= y^1)$ and $y(= y^2)$ are Cartesian coordinates in the plane. Let $\{\xi^i\} = \{\phi, \psi, y^3\}$, where $\phi = \text{Re}\, g$ and $\psi = \text{Im}\, g$. Consider the example $g(z) = z^2$ and compute $\{\mathbf{g}_i\}$, $\{\mathbf{g}^i\}$, g_{ij}, g^{ij} in terms of the y^i. Sketch the coordinate surfaces, i.e., the surfaces on which each coordinate is constant.

Problem 3.6 Consider an oblique coordinate system $\{\xi^i\}$ for which $\mathbf{g}_1 = \mathbf{i}_1$, $\mathbf{g}_2 = \cos\phi \mathbf{i}_2 + \sin\phi \mathbf{i}_1$, $\mathbf{g}_3 = \mathbf{i}_3$ with ϕ a constant angle.

(a) Find the y^i as functions of the ξ^i. (Hint: construct the position vector and suppose the ξ^i vanish at the origin.) Invert these relations.
(b) Obtain g_{ij}, g^{ij} and find $\{\mathbf{g}^i\}$ in terms of $\{\mathbf{i}_i\}$.
(c) If $v_i^{(c)}$ are the Cartesian components of a vector, what are the co- and contravariant components of the vector in the $\{\xi^i\}$ system?

3.2 Second-order tensors

Vectors are essentially *first*-order tensors in the sense that they are fully specified by components, relative to a basis, bearing a *single* index (which, of course, ranges over the dimension of the space). That is, they have 3^1 components, where the exponent is the order and the base is the dimension of the space. Second-order tensors, with 3^2 components, are effectively 9-dimensional vectors. They are linear combinations of the elements of $\{\mathbf{g}_i \otimes \mathbf{g}_j\}$, $\{\mathbf{g}_i \otimes \mathbf{g}^j\}$, $\{\mathbf{g}^i \otimes \mathbf{g}_j\}$, or $\{\mathbf{g}^i \otimes \mathbf{g}^j\}$, and their components accordingly bear two indices. To demonstrate the linear independence of the elements of the first set, we equate to the zero tensor (i.e., the 9-dimensional zero vector) the linear combination $\alpha^{ij}\mathbf{g}_i \otimes \mathbf{g}_j$. Now operate the result on $\mathbf{g}^k$ to obtain $\alpha^{ik}\mathbf{g}_i = \mathbf{0}$ and thus conclude that all α^{ij} vanish. Because the set $\{\mathbf{g}_i \otimes \mathbf{g}_j\}$ contains nine linearly independent elements, it constitutes a basis for the linear space of second-order tensors. That is, any such tensor, $\mathbf{A}$ say, is expressible as

$$\mathbf{A} = A^{ij}\mathbf{g}_i \otimes \mathbf{g}_j, \tag{3.39}$$

with unique coefficients A^{ij}. Reasoning in precisely the same way, we have the alternative representations

$$\mathbf{A} = A_{ij}\mathbf{g}^i \otimes \mathbf{g}^j = A^i_{\cdot j}\mathbf{g}_i \otimes \mathbf{g}^j = A_i^{\cdot j}\mathbf{g}^i \otimes \mathbf{g}_j, \tag{3.40}$$

again with unique coefficients. The A_{ij} and A^{ij} respectively are the *co-* and *contravariant* components, and the $A^i_{\cdot j}$ and $A_i^{\cdot j}$ respectively are the *left-mixed* and *right-mixed* components; the dot is inserted as a device to distinguish one type from the other.

With (3.40) in hand we have

$$\mathbf{A}\mathbf{g}^j = (A^{kl}\mathbf{g}_k \otimes \mathbf{g}_l)\mathbf{g}^j = A^{kl}\mathbf{g}_k(\mathbf{g}_l \cdot \mathbf{g}^j) = A^{kl}\delta_l^j\mathbf{g}_k = A^{kj}\mathbf{g}_k, \tag{3.41}$$

and dotting with $\mathbf{g}^i$ gives

$$A^{ij} = \mathbf{g}^i \cdot \mathbf{A}\mathbf{g}^j = \mathbf{A} \cdot \mathbf{g}^i \otimes \mathbf{g}^j. \tag{3.42}$$

Thus, the contravariant components are simply the projections of the tensor onto the corresponding reciprocal basis elements. In the same way we can easily establish the similar formulas

$$A_{ij} = \mathbf{A} \cdot \mathbf{g}_i \otimes \mathbf{g}_j, \quad A^i_{\cdot j} = \mathbf{A} \cdot \mathbf{g}^i \otimes \mathbf{g}_j, \quad \text{and} \quad A_i^{\cdot j} = \mathbf{A} \cdot \mathbf{g}_i \otimes \mathbf{g}^j. \tag{3.43}$$

Beyond this, we can use the metric and reciprocal metric to connect the various sets of components via index raising and lowering. For example,

$$g^{lm}A_{kl}\mathbf{g}^k = A_{kl}\mathbf{g}^k(\mathbf{g}^l \cdot \mathbf{g}^m) = \mathbf{A}\mathbf{g}^m = A^{ij}\mathbf{g}_i(\mathbf{g}_j \cdot \mathbf{g}^m) = A^{im}\mathbf{g}_i, \tag{3.44}$$

and a further dotting with $\mathbf{g}^n$ gives

$$A^{nm} = g^{nk}g^{ml}A_{kl}. \tag{3.45}$$

In the same way,

$$A^{ij} = g^{jk}A_k^{\cdot j} = g^{jk}A^i_{\cdot k}, \quad A_{ij} = g_{ik}g_{jl}A^{kl}, \quad \text{etc.} \tag{3.46}$$

Problem 3.7 Show that $\mathbf{A}^t = A^i_{\cdot j}\mathbf{g}^j \otimes \mathbf{g}_i$, and hence that $A^i_{\cdot j} = A_j^{\cdot i}$ if $\mathbf{A}$ is symmetric. For symmetric $\mathbf{A}$, show that $A_{ij} = A_{ji}$ and $A^{ij} = A^{ji}$. What if $\mathbf{A}$ is skew-symmetric, i.e., $\mathbf{A}^t = -\mathbf{A}$?

The unit tensor $\mathbf{i}$ is defined by $\mathbf{i}\mathbf{v} = \mathbf{v}$ for any vector $\mathbf{v}$. Accordingly, it has the various representations

$$\begin{aligned}
\mathbf{i} &= (\mathbf{g}_i \cdot \mathbf{i}\mathbf{g}_j)\mathbf{g}^i \otimes \mathbf{g}^j \\
&= (\mathbf{g}_i \cdot \mathbf{g}_j)\mathbf{g}^i \otimes \mathbf{g}^j = g_{ij}\mathbf{g}^i \otimes \mathbf{g}^j = \mathbf{g}_j \otimes \mathbf{g}^j \\
&= \delta^i_j\mathbf{g}_i \otimes \mathbf{g}^j = g^{ij}\mathbf{g}_i \otimes \mathbf{g}_j = \mathbf{g}^j \otimes \mathbf{g}_j = \delta_i^j\mathbf{g}^i \otimes \mathbf{g}_j.
\end{aligned} \tag{3.47}$$

The metric and reciprocal metric, respectively, are therefore the co- and contravariant components of the unit tensor, whereas the left- and right-mixed components coincide and are equal to the Kronecker delta. Note that the mixed components are the same in every coordinate system, whereas the co- and contravariant components vary from one system to another. Obviously the unit tensor is symmetric.

This brings us to the central property of tensors, namely their coordinate invariance. We can think in terms of imposing such invariance as part of the definition of a tensor. Given this, it is obvious that since the bases used to represent tensors depend on the coordinates, the components of these tensors must compensate to ensure that the tensors themselves do not. To explore this issue, consider a coordinate transformation from $\{\xi^i\}$ to $\{\bar{\xi}^i\}$, say, given by the functions $\bar{\xi}^i = f^i(\xi^j)$. We suppose these functions to be differentiable and invertible, at least locally. For example, we can assume a locally invertible relation between the $\bar{\xi}^i$ and Cartesian coordinates y^i in the same way that we previously assumed such a relation between the latter and the ξ^i. This, in turn, ensures the local invertibility of the functions f^i. Using the $\bar{\xi}^i$ we can construct the elements $\bar{\mathbf{g}}_i$ of the induced tangent basis and the elements $\bar{\mathbf{g}}^i$ of the associated reciprocal basis. Naturally, this is accomplished by using (3.2) and (3.11), but with the ξ^i replaced by $\bar{\xi}^i$. The chain rule then furnishes

$$\bar{\mathbf{g}}_i = \frac{\partial \xi^j}{\partial \bar{\xi}^i}\mathbf{g}_j \quad \text{and} \quad \bar{\mathbf{g}}^i = \frac{\partial \bar{\xi}^i}{\partial \xi^j}\mathbf{g}^j, \tag{3.48}$$

where, in a harmless abuse of notation, we have replaced the functions f^i by their values. These are the change-of-basis formulas referred to previously. Then for any vector $\mathbf{v}$,

$$v^i\mathbf{g}_i = \mathbf{v} = \bar{v}^j\bar{\mathbf{g}}_j = \bar{v}^j\frac{\partial \xi^i}{\partial \bar{\xi}^j}\mathbf{g}_i, \quad \text{and thus} \quad v^i = \frac{\partial \xi^i}{\partial \bar{\xi}^j}\bar{v}^j, \tag{3.49}$$

and, in the same way,

$$v_i = \frac{\partial \bar{\xi}^j}{\partial \xi^i}\bar{v}_j. \tag{3.50}$$

The invertibility of the coordinate transformation implies that the same results hold with the barred and unbarred variables interchanged, i.e.,

$$\bar{v}^i = \frac{\partial \bar{\xi}^i}{\partial \xi^j}v^j \quad \text{and} \quad \bar{v}_i = \frac{\partial \xi^j}{\partial \bar{\xi}^i}v_j. \tag{3.51}$$

Exactly the same line of reasoning leads to the transformation rules

$$\bar{A}_{ij} = \frac{\partial \xi^k}{\partial \bar{\xi}^i}\frac{\partial \xi^l}{\partial \bar{\xi}^j}A_{kl}, \quad \bar{A}^{ij} = \frac{\partial \bar{\xi}^i}{\partial \xi^k}\frac{\partial \bar{\xi}^j}{\partial \xi^l}A^{kl}, \quad \text{etc.,} \tag{3.52}$$

for the components of second-order tensors.

3.3 Derivatives and connections

We've already introduced and used the gradient operation on scalars defined by (3.7). We extend this to vector-valued functions $\mathbf{v}(\mathbf{y})$ and define the gradient, *grad*$\mathbf{v}$, to be the second-order tensor such that

$$d\mathbf{v} = (grad\mathbf{v})d\mathbf{y}. \tag{3.53}$$

Regarding $\mathbf{v}$ as a function of the coordinates via $\mathbf{v}(\mathbf{y}(\xi^i))$, we have $d\mathbf{v} = \mathbf{v}_{,i}d\xi^i = \mathbf{v}_{,i}(\mathbf{g}^i \cdot d\mathbf{y})$. Inserting this on the left-hand side of the definition and demanding that equality hold for all $d\mathbf{y}$ gives

$$grad\mathbf{v} = \mathbf{v}_{,i} \otimes \mathbf{g}^i. \tag{3.54}$$

To extract a representation in terms of components, we write

$$\mathbf{v}_{,i} = (v^j\mathbf{g}_j)_{,i} = v^j_{,i}\mathbf{g}_j + v^j\mathbf{g}_{j,i}, \tag{3.55}$$

where we account for the fact that the $\mathbf{g}_j$, unlike their Cartesian counterparts, generally depend on the coordinates. Now $\mathbf{g}_{j,i}$, like any other vector, may be represented in terms of a basis. We choose the basis $\{\mathbf{g}_k\}$ at the particular point in question, with coordinates ξ^i, and use (3.33) and (3.34) to conclude that

$$\mathbf{g}_{j,i} = \Gamma^k_{ji}\mathbf{g}_k, \quad \text{where} \quad \Gamma^k_{ji} = \mathbf{g}^k \cdot \mathbf{g}_{j,i}. \tag{3.56}$$

The gammas are called the *connection* coefficients; they effectively connect the *tangent space Span*$\{\mathbf{g}_k\}$ at the point ξ^i to the tangent space at the neighboring point $\xi^i + d\xi^i$.

We've been quite cavalier about the choices of bases used in our various representations. In Euclidean space we could just as well have used any other basis; for example, the basis $\{\mathbf{g}_k\}$ existing at some remote point of the space. The reason is that Euclidean space is effectively flat in the sense that it coincides with its own tangent space. Accordingly, vectors attached to points remote from each other can be added with impunity. In a more general space, or *manifold*, vectors at a particular point are elements of the space tangent to the manifold at that point. Vectors attached to distinct points of the manifold belong to distinct tangent spaces, and their addition would thus have no vectorial meaning. The choice we made in the course of arriving at (3.56) thus makes sense in spaces more general than Euclidean, as we shall see later.

However, in Euclidean space we can invoke (3.2) and, assuming the position field $\mathbf{y}(\xi^i)$ to be twice differentiable, conclude that

$$\Gamma^k_{ji} = \mathbf{g}^k \cdot \mathbf{y}_{,ji} = \mathbf{g}^k \cdot \mathbf{y}_{,ij} = \mathbf{g}^k \cdot \mathbf{g}_{i,j} = \Gamma^k_{ij}, \tag{3.57}$$

where the comma followed by two subscripts is a shorthand for second partial derivatives. Accordingly, the gammas are symmetric in the subscripts.

Problem 3.8 Compute the gammas for spherical coordinates.

It turns out that in a certain non-Euclidean space of relevance to plasticity theory the gammas do not possess this symmetry. To cover this possibility we define the *torsion*

$$T_{\cdot ij}^{\,k} = \Gamma^k_{[ij]}, \tag{3.58}$$

where the square bracket is used to denote skew symmetrization; that is,

$$\Gamma^k_{[ij]} = \frac{1}{2}(\Gamma^k_{ij} - \Gamma^k_{ji}). \tag{3.59}$$

Clearly, then, the torsion vanishes in Euclidean geometry. Later, we will encounter another non-Euclidean geometry wherein the torsion also vanishes. This is *Riemannian* geometry, which, in the extension to four dimensions, furnishes the setting for Einstein's astounding theory of gravitation.

Indeed in a more general geometry one does not have (3.56), but rather

$$d\mathbf{g}_j = \Gamma^k_{ji}\mathbf{g}_k d\xi^i, \tag{3.60}$$

by which we mean

$$\dot{\mathbf{g}}_j = \Gamma^k_{ji}\mathbf{g}_k \dot{\xi}^i, \tag{3.61}$$

where $\xi^i(t)$ are the parametric equations of a curve in the manifold, superposed dots are derivatives with respect to t, and all quantities are defined on the curve. See Section 9.7 of the book by Stoker, or the delightfully lucid discussion in Chapter 5 of Brillouin's book. Given the Γ^k_{ji}, we have a linear system of ordinary differential equations, and a standard existence-uniqueness theorem ensures that these can be integrated, granted the availability of initial conditions $\mathbf{g}_j(t_0)$, giving $\mathbf{g}_j(t)$. Such geometries, which subsume all those of interest to us, are said to be *affinely connected* because the $\dot{\xi}^i$ enter (3.61) linearly. We can think of (3.61) as defining the directional derivatives of the $\mathbf{g}_j$ in the direction of the tangent, with components $\dot{\xi}^i$, to the curve. The gammas appearing in the expressions for these derivatives go by the name *Koszul connection* in the recent literature. See the book by Szerkeres, for example.

On the other hand, (3.56) makes sense only if the $\mathbf{g}_j$ and Γ^k_{ji} are defined on the manifold in a neighborhood of the point on the curve in question. Indeed if this is the case, then by the chain rule the left-hand side of (3.61) becomes $\mathbf{g}_{j,i}\dot{\xi}^i$, provided that the $\mathbf{g}_j$ are differentiable, and (3.56) follows if the resulting equation is to hold for all curves; that is, for arbitrary $\dot{\xi}^i$. Following Einstein, we then have a space with *distant parallelism*. These are exemplified by Euclidean space, naturally, and also by a certain space with torsion alluded to previously that will prove to be of interest to us. Clearly, then, the passage from (3.61) to (3.56) is not automatic, but requires that certain integrability conditions be satisfied. For example, if the $\mathbf{g}_j$ are twice differentiable, so that $\mathbf{g}_{j,[kl]}$ vanishes, then for (3.56) to hold it is necessary—and in a simply connected region, also sufficient (see Ciarlet's book)—that

$$(\Gamma^i_{jk}\mathbf{g}_i)_{,l} = (\Gamma^i_{jl}\mathbf{g}_i)_{,k}. \tag{3.62}$$

It turns out that this imposes a serious restriction on the gammas and their derivatives. This restriction is automatically satisfied in Euclidean space provided that the position field therein is thrice differentiable. For we then have $\mathbf{g}_{j,kl} = \mathbf{y}_{,jkl}$, which is completely symmetric in the subscripts.

We have digressed, perhaps too far, to discuss non-Euclidean spaces. Further discussion will be forthcoming as the need arises. In the meantime we return to the lowly gradient of a vector field in Euclidean space. Thus, we combine (3.55) and (3.56) and shuffle the indices to obtain

$$\mathbf{v}_{,i} = v^k_{;i}\mathbf{g}_k, \quad \text{where} \quad v^k_{;i} = v^k_{,i} + \Gamma^k_{ji}v^j \tag{3.63}$$

is the so-called *covariant derivative*. We then have

$$grad\mathbf{v} = v^k_{;i}\mathbf{g}_k \otimes \mathbf{g}^i, \tag{3.64}$$

and hence the interpretation of the $v^k_{;i}$ as the left-mixed components of $grad\mathbf{v}$. Given these, we can raise and lower indices in the usual way to obtain the various other kinds of components.

To arrive at (3.64) we started with the tangent-basis decomposition of $\mathbf{v}$ in terms of its contravariant components. Of course we might have used the alternative reciprocal-basis decomposition, in terms of covariant components, instead. Then, in place of (3.55) we would have had

$$\mathbf{v}_{,i} = v_{j,i}\mathbf{g}^j + v_j\mathbf{g}^j_{,i}. \tag{3.65}$$

This time we use the first of (3.33) with (3.35) to express the $\mathbf{g}^j_{,i}$ as

$$\mathbf{g}^j_{,i} = (\mathbf{g}_k \cdot \mathbf{g}^j_{,i})\mathbf{g}^k, \tag{3.66}$$

in terms of the reciprocal basis existing at the point in question, in which the coefficients follow by differentiating (3.15) and substituting (3.56), giving

$$\mathbf{g}_k \cdot \mathbf{g}^j_{,i} = -\Gamma^j_{ki}. \tag{3.67}$$

Thus,

$$\mathbf{v}_{,i} = v_{k;i}\mathbf{g}^k, \quad \text{where} \quad v_{k;i} = v_{k,i} - \Gamma^j_{ki}v_j. \tag{3.68}$$

This is also called a covariant derivative. Note that in both types the second of the lower indices on the gammas corresponds to the index used to label the derivative. Be aware that some workers use a definition of the gammas in which the lower indices are switched relative to our definition. In any case we now have

$$grad\mathbf{v} = v_{k;i}\mathbf{g}^k \otimes \mathbf{g}^i, \tag{3.69}$$

and hence the interpretation of $v_{k;i}$ as the covariant components of the (same) tensor $grad\mathbf{v}$.

We note in passing that the covariant derivative operations, applied directly to the tangent and reciprocal basis elements, yield

$$\mathbf{g}_{i;j} = \mathbf{0} \quad \text{and} \quad \mathbf{g}^i_{;j} = \mathbf{0}, \tag{3.70}$$

i.e., the basis elements are covariantly constant. These extend their Cartesian counterparts; namely, $\partial\mathbf{i}_i/\partial y^j = \mathbf{0}$, to arbitrary coordinates.

With the gradient in hand we can compute the *divergence*, $div\mathbf{v}$, of a vector field. This is the scalar field defined intrinsically, i.e., without reference to coordinates, by

$$div\mathbf{v} = tr(grad\mathbf{v}). \tag{3.71}$$

Using (3.54) and (3.63) we have

$$div\mathbf{v} = \mathbf{v}_{,i} \cdot \mathbf{g}^i = v^k_{;i}\mathbf{g}_k \cdot \mathbf{g}^i = v^i_{;i} = (g^{ki}v_k)_{;i}. \tag{3.72}$$

Alternatively, using (3.69),

$$div\mathbf{v} = v_{k;i}\mathbf{g}^k \cdot \mathbf{g}^i = g^{ki}v_{k;i}. \tag{3.73}$$

If these are to agree, as they must, then it seems the g^{ki} can be taken outside the covariant derivative. We return to this point below, after we introduce covariant derivatives of second-order tensors.

Another extremely important differential operator is the Laplacian, $\Delta\phi$, of a scalar-valued function ϕ, defined intrinsically by

$$\Delta\phi = div(grad\phi), \tag{3.74}$$

where *grad* is the gradient. Combining this with (3.54) and (3.11) results in

$$\begin{aligned} \Delta\phi &= (grad\phi)_{,i} \cdot \mathbf{g}^i = \mathbf{g}^i \cdot (\phi_{,j}\mathbf{g}^j)_{,i} \\ &= \mathbf{g}^i \cdot (\phi_{,ji}\mathbf{g}^j + \phi_{,j}\mathbf{g}^j_{,i}) = g^{ki}\phi_{;ki}, \end{aligned} \tag{3.75}$$

where

$$\phi_{;ki} = \phi_{,ki} - \Gamma^j_{ki}\phi_{,j}. \tag{3.76}$$

Note that $\phi_{;ki} = \phi_{;ik}$ in a torsion-free space, provided that ϕ is twice differentiable.

Moving on to second-order tensors, we proceed, as in Chapter 1, to define the divergence of a tensor field $\mathbf{A}(\mathbf{y})$ intrinsically, at a fixed point of space, to be the vector field $div\mathbf{A}$ such that

$$\mathbf{c} \cdot div\mathbf{A} = div(\mathbf{A}^t\mathbf{c}), \tag{3.77}$$

for any *fixed* vector $\mathbf{c}$. Writing $\mathbf{v}$ for $\mathbf{A}^t\mathbf{c}$ and invoking (3.72), we then have

$$\mathbf{c} \cdot div\mathbf{A} = \mathbf{g}^i \cdot \mathbf{A}^t_{,i}\mathbf{c} = (\mathbf{A}^t_{,i})^t\mathbf{g}^i \cdot \mathbf{c}. \tag{3.78}$$

We have rather nonchalantly failed to specify which comes first: the transpose or the derivative. In fact, the order doesn't matter.

Problem 3.9 Prove this.

Then,

$$\mathbf{c} \cdot [div\mathbf{A} - (\mathbf{A}_{,i})\mathbf{g}^i] = 0, \tag{3.79}$$

and as this purports to hold for all fixed $\mathbf{c}$, we may take it to coincide with the square bracket, fixed at the point in question, to conclude that the norm of the bracket, and hence the bracket itself, vanishes. Thus,

$$div\mathbf{A} = (\mathbf{A}_{,j})\mathbf{g}^j. \tag{3.80}$$

This exceedingly useful formula conceals fairly complicated component expressions. For example, starting with the representation (3.39), differentiating everything in sight and applying the product rule to the tensor products, we get

$$\mathbf{A}_{,j} = A^{ik}_{;j}\mathbf{g}_i \otimes \mathbf{g}_k, \tag{3.81}$$

where

$$A^{ik}_{;j} = A^{ik}_{,j} + A^{il}\Gamma^k_{lj} + A^{lk}\Gamma^i_{lj} \tag{3.82}$$

is one among four kinds of covariant derivative. Another kind follows by using the first of Eqs. (3.40): Thus,

$$\mathbf{A}_{,j} = A_{ik;j}\mathbf{g}^i \otimes \mathbf{g}^k, \tag{3.83}$$

where

$$A_{ik;j} = A_{ik,j} - A_{lk}\Gamma^l_{ij} - A_{il}\Gamma^l_{kj}. \tag{3.84}$$

Finally,

$$divA = A^{ij}_{;j}\mathbf{g}_i = A_{ik;j}g^{kj}\mathbf{g}^i, \quad \text{etc.} \tag{3.85}$$

Problem 3.10 Show that for two differentiable vector fields **u** and **v**,

$$(\mathbf{u} \otimes \mathbf{v})_{,i} = \mathbf{u}_{,i} \otimes \mathbf{v} + \mathbf{u} \otimes \mathbf{v}_{,i}.$$

Problem 3.11 Verify (3.82) and (3.84), and write out expressions for the remaining two kinds of covariant derivative, namely $A^{i}_{\cdot k;j}$ and $A_{k;j}^{\cdot i}$.

The covariant derivative provides a covariant (i.e., tensorially meaningful) extension of the partial derivative. Unsurprisingly, it possesses all of the useful properties of the latter, including the product rule. In fact, modern treatments include the product rule among the defining properties of the covariant derivative. Here we content ourselves with a verification of the product rule in the example

$$\mathbf{v} = \mathbf{A}\mathbf{u} = (A^{ij}\mathbf{g}_i \otimes \mathbf{g}_j)\mathbf{u} = A^{ij}u_j\mathbf{g}_i; \quad \text{that is,} \quad v^i = A^{ij}u_j. \tag{3.86}$$

Then,

$$\begin{aligned} v^i_{;k} &= v^i_{,k} + v^j\Gamma^i_{jk} = A^{ij}_{,k}u_j + A^{ij}u_{j,k} + A^{jl}u_l\Gamma^i_{jk} \\ &= A^{ij}_{,k}u_j + A^{ij}u_{j,k} + A^{lj}\Gamma^i_{lk}u_j + (A^{il}\Gamma^j_{lk}u_j - A^{ij}\Gamma^l_{jk}u_l) \\ &= (A^{ij}_{,k} + A^{il}\Gamma^j_{lk} + A^{lj}\Gamma^i_{lk})u_j + A^{ij}(u_{j,k} - \Gamma^l_{jk}u_l) \\ &= A^{ij}_{;k}u_j + A^{ij}u_{j;k}, \end{aligned} \tag{3.87}$$

in which zero has been added in the second line, in the guise of the parenthetical term.

In the same way we can establish that

$$(A^{ij}B_{kl})_{;m} = A^{ij}{}_{;m}B_{kl} + A^{ij}B_{kl;m} \tag{3.88}$$

and

$$(\phi v^k)_{;l} = \phi_{;l}v^k + \phi v^k_{;l}, \tag{3.89}$$

etc., with the proviso that the covariant derivative of a scalar field is the same as its partial derivative, i.e.,

$$\phi_{;l} = \phi_{,l}. \tag{3.90}$$

We extend this convention to any *invariant*. Thus, for vectors,

$$\mathbf{v}_{;k} = \mathbf{v}_{,k} \tag{3.91}$$

and for second-order tensors,

$$\mathbf{A}_{;k} = \mathbf{A}_{,k}. \tag{3.92}$$

For example,

$$\mathbf{A}_{,k} = A^{ij}_{;k}\mathbf{g}_i \otimes \mathbf{g}_j + A^{ij}(\mathbf{g}_{i;k} \otimes \mathbf{g}_j + \mathbf{g}_i \otimes \mathbf{g}_{j;k}), \tag{3.93}$$

and (3.81) follows from (3.70).

Concerning the metric, we combine (3.21) and (3.56), obtaining

$$g_{ij,k} = (\mathbf{g}_i \cdot \mathbf{g}_j)_{,k} = g_{lj}\Gamma^l_{ik} + g_{li}\Gamma^l_{jk}; \quad \text{that is,} \quad g_{ij;k} = 0. \tag{3.94}$$

This establishes that the unit tensor (3.47) is spatially uniform, i.e., independent of the coordinates. A more clever derivation follows by invoking the product rule for covariant derivatives together with (3.70). Thus,

$$g_{ij;k} = (\mathbf{g}_i \cdot \mathbf{g}_j)_{;k} = \mathbf{g}_{i;k} \cdot \mathbf{g}_j + \mathbf{g}_i \cdot \mathbf{g}_{j;k}, \tag{3.95}$$

which of course vanishes identically. In the same way we easily establish that

$$g^{ij}_{;k} = 0. \tag{3.96}$$

The metric and reciprocal metric are therefore covariantly constant, and the connection is said to be *metric-compatible.*

3.4 The Levi-Civita connection

This important result, known as *Ricci's lemma*, explains why it is indeed correct to take the metric outside the covariant derivative, as in the comparison of (3.72) and (3.73). It leads straightforwardly to an important relation giving the gammas in terms of the metric and its partial derivatives. To derive this we first introduce the auxiliary connection symbols

$$\Gamma_{jik} = g_{lj}\Gamma^l_{ik}, \tag{3.97}$$

allowing (3.94) to be expressed more tidily as

$$g_{ij,k} = \Gamma_{jik} + \Gamma_{ijk}. \tag{3.98}$$

Note that $\Gamma_{j[ik]} = g_{lj}T^l_{\cdot ik}$, so that $\Gamma_{jik} = \Gamma_{jki}$ in a torsionless (e.g., Euclidean or Riemannian) geometry. It follows from (3.98) that

$$\frac{1}{2}(g_{ki,j} + g_{kj,i} - g_{ij,k}) = \Gamma_{i[kj]} + \Gamma_{j[ki]} + \Gamma_{k(ij)}, \tag{3.99}$$

where the round braces are used to denote symmetrization, i.e.,

$$\Gamma_{k(ij)} = \frac{1}{2}(\Gamma_{kij} + \Gamma_{kji}). \tag{3.100}$$

Accordingly, in the torsionless Euclidean space with which we are currently concerned,

$$\Gamma_{kij} = \frac{1}{2}(g_{ki,j} + g_{kj,i} - g_{ij,k}). \tag{3.101}$$

To recover the connection symbols we use

$$\Gamma^m_{ij} = \delta^m_l \Gamma^l_{ij} = g^{mk} g_{lk} \Gamma^l_{ij} = g^{mk} \Gamma_{kij}, \tag{3.102}$$

arriving finally at

$$\Gamma^m_{ij} = \frac{1}{2} g^{mk}(g_{ki,j} + g_{kj,i} - g_{ij,k}). \tag{3.103}$$

This is called the *Levi-Civita* connection. Note that its derivation, from (3.98), made no use of (3.56), but instead relied only on the notion of a metric and a metric-compatible connection. Accordingly, it applies equally to Riemannian geometry, defined simply as a space endowed with a (symmetric) metric and a Levi-Civita connection. In particular, then, Euclidean geometry is a special case.

In fact, we are following the program of Cartan's intuitive approach to Riemannian geometry, whereby a Euclidean tangent space is associated with a point of the Riemannian manifold. In this tangent space the dot product is well defined, and may be used to generate a metric as in (3.21). The metric of the Riemannian space is then defined to be that of the Euclidean tangent space *at* the point in question. In fact, we can go a step further and take the Levi-Civita connection of the Riemannian space to coincide with that of the Euclidean tangent space, again *at* the point in question. We then speak of an *osculating* Euclidean space. Thus, the partial derivatives of the metrics are also matched and the Euclidean rules for covariant differentiation are preserved in the Riemannian geometry. Concise descriptions of this procedure are given in the books by Willmore, Lichnerowicz, Stoker, and Cartan. In this way much of Euclidean geometry is neatly carried over to Riemannian geometry. However, the curvature of Riemannian space, a concept to be introduced shortly, intervenes to prevent the matching of higher-order derivatives of the metric. In a space with torsion we can also construct a Euclidean tangent space but this cannot be osculating because in the latter the connection is symmetric.

By the way, the so-called non-metric geometry conceived by Weyl, in which the connection is not metric compatible, is also of interest. In this geometry Ricci's lemma is replaced by $g_{ij;k} = q_{ijk}$ (with $q_{[ij]k} = 0$), where q_{ijk}, the source of non-metricity, may be interpreted as a continuous distribution of point defects, such as interstitial atoms or vacancies, in a crystal. See the paper by Povstenko, for example. We are not concerned with such geometries here, however.

A most useful property of the Levi-Civita connection follows by contracting (3.103) on the upper index. Using the symmetry of the metric, we find that

$$2\Gamma^i_{ij} = g^{ik}g_{ik,j} + (g^{ik}g_{kj,i} - g^{ki}g_{ij,k}), \tag{3.104}$$

in which the second parenthetical term is seen, on relabelling the summation indices, to be the same as the first. Accordingly, the parenthesis vanishes, leaving

$$\Gamma^i_{ij} = \frac{1}{2}g^{ik}g_{ik,j}. \tag{3.105}$$

This is simplified by recalling that the derivative of the determinant of a matrix with respect to one of its components is the corresponding component of the cofactor matrix. Thus,

$$\frac{\partial g}{\partial g_{ik}} = gg^{ik}, \tag{3.106}$$

so that

$$\Gamma^i_{ij} = \frac{1}{2g}\frac{\partial g}{\partial g_{ik}}g_{ik,j} = \frac{1}{2g}g_{,j} = (\sqrt{g})_{,j}/\sqrt{g}. \tag{3.107}$$

With this result in hand let us revisit the divergence and Laplacian operators. For the first of these, (3.63) and (3.72) give

$$div\mathbf{v} = v^i_{;i} = v^i_{,i} + \frac{(\sqrt{g})_{,j}}{\sqrt{g}}v^j = \frac{1}{\sqrt{g}}(\sqrt{g}v^i)_{,i}, \tag{3.108}$$

which does not require the explicit (and rather tedious) computation of the connection coefficients. For the vector field $\mathbf{v} = grad\phi$ we have $v^i = g^{ij}\phi_{,j}$; Eqs. (3.74) and (3.108) then furnish the Laplacian

$$\Delta\phi = \frac{1}{\sqrt{g}}(\sqrt{g}g^{ij}\phi_{,j})_{,i}. \tag{3.109}$$

Problem 3.12 Write out the Laplacian $\Delta f(\xi, \eta, \phi)$ explicitly in terms of f and its partial derivatives with respect to ξ, η, ϕ for the coordinate systems of Problem 3.4.

Problem 3.13 In cylindrical polar coordinates $\{r, \theta, z\}$ the position function is given by $\mathbf{y} = r\mathbf{e}_r(\theta) + z\mathbf{i}_3$, where r is the radius from the z-axis, z measures distance along this axis, θ is the azimuthal angle, and $\mathbf{e}_r(\theta) = \cos\theta\mathbf{i}_1 + \sin\theta\mathbf{i}_2$. In domains with cone-shaped boundaries, it may be more convenient to use a *conical-polar* coordinate system. For this purpose we introduce coordinates $\{\bar{r}, \theta, \bar{z}\}$, where $r = \bar{r}\cos\phi$,

$\bar{z} = z - \bar{r}\sin\phi$, and ϕ is the *constant* cone angle. The position function may then be written $\mathbf{y} = \bar{r}\bar{\mathbf{e}}_r(\theta) + \bar{z}\mathbf{i}_3$, where $\bar{\mathbf{e}}_r(\theta) = \cos\phi\mathbf{e}_r(\theta) + \sin\phi\mathbf{i}_3$ (draw a figure). Let $\{\xi^i\} = \{\bar{r}, \theta, \bar{z}\}$ and write out the divergence $A^{ij}_{;j}$ explicitly in terms of A^{ij} and their partial derivatives.

Problem 3.14 Show that the Laplacian of a function $f(r, \theta, \phi)$ in spherical coordinates is

$$\Delta f = \frac{1}{r^2}[(r^2 f_r)_{,r} + (\sec^2\phi)f_{\theta\theta} + \sec\phi(f_\phi \cos\phi)_\phi],$$

where the subscripts stand for partial derivatives.

Problem 3.15 Position in a *helical* coordinate system $\{\xi^i\} = \{r, \theta, s\}$ is described by

$$\mathbf{y}(\xi^i) = \mathbf{r}(s) + r\cos\theta\mathbf{n}(s) + r\sin\theta\mathbf{b}(s),$$

where $\mathbf{r}(s)$ is the equation of the center-line of a helical tube, s is arclength along the center-line, $\mathbf{n}(s)$ is the *principal normal* to the center-line, and $\mathbf{b}(s)$ is the *binormal.* The coordinates r and θ, respectively, are the radius from the center-line and the counterclockwise azimuth measured from the principal normal in a plane cross section of the tube. The curvature of the center-line is $\kappa(s) = |\mathbf{t}'(s)|$, where $\mathbf{t} = \mathbf{r}'(s)$ is the unit tangent to the center-line. Wherever κ is non-zero, we stipulate that $\mathbf{n} = \kappa^{-1}\mathbf{t}'(s)$ and $\mathbf{b} = \mathbf{t} \times \mathbf{n}$ so that $\{\mathbf{t},\mathbf{n},\mathbf{b}\}$ is an orthonormal basis at every point of the curve.

The Serret–Frenet equations for smooth space curves are (see any elementary geometry or dynamics text).

$$\mathbf{n}'(s) = \tau\mathbf{b} - \kappa\mathbf{t} \quad \text{and} \quad \mathbf{b}'(s) = -\tau\mathbf{n},$$

where $\tau(s)$ is the *torsion* of the center-line (not to be confused with the torsion, i.e., the skew part of the connection, referred to in the text).

Let $\{\xi^i\} = \{r, \theta, s\}$ and obtain $\{\mathbf{g}_i\}$, $\{\mathbf{g}^i\}$, g_{ij}, g^{ij} in terms of $\mathbf{t},\mathbf{n},\mathbf{b},\kappa,\tau$ and the coordinates. Show that this coordinate system is *non-orthogonal*; i.e., the matrix (g_{ij}) is not diagonal.

Helices of uniform radius and pitch are distinguished by the property that the curvature and torsion are constants, independent of arclength. Obtain an explicit expression for the Laplacian $\Delta f(r, \theta)$ in this special case, for a function f that is independent of arclength. This problem is of obvious importance in diverse applications including, for example, heat conduction in helical wires and fluid flow in helical tubes.

Problem 3.16 Work out the transformation formula

$$\bar{A}^{k}_{\cdot ij} = \frac{\partial \bar{\xi}^k}{\partial \xi^l} \frac{\partial \xi^m}{\partial \bar{\xi}^i} \frac{\partial \xi^n}{\partial \bar{\xi}^j} A^{l}_{\cdot mn}$$

for components $A^{k}_{\cdot ij}$ of a third-order tensor.

(a) Thus, show that the connection coefficients Γ^k_{ij} are *not* the components of a tensor, firstly by proceeding from the formula (3.56) for the connection coefficients, and secondly by proceeding directly from the Levi-Civita form (3.103) without using (3.56). In fact this result follows from the tensorial property of the covariant derivatives of a tensor, whether or not the connection has either of the forms (3.56) or (3.103). See the book by Lovelock and Rund or Szekeres' book for fuller discussions.

(b) Suppose the connection is asymmetric, with a non-zero torsion $T^{k}_{\cdot ij} = \Gamma^k_{[ij]}$. Show that the $T^{k}_{\cdot ij}$ *are* the components of a tensor, provided that the coordinates $\bar{\xi}^i$ are twice-differentiable functions of the ξ^i.

(c) Using (3.16), argue that $\epsilon_{ijk;l} = 0$ and $\epsilon^{ijk}_{;l} = 0$ *without* invoking the formulae for the covariant derivatives of third-order tensor components. Use a similar argument to explain why the metric and reciprocal metric are covariantly constant without resorting to Ricci's lemma.

3.5 The curl, Stokes' theorem, and curvature

The last of the main differential operators in Euclidean space is the curl, defined, for a vector field $\mathbf{v}$, to be the vector *curl*$\mathbf{v}$ such that

$$\mathbf{c} \cdot curl\mathbf{v} = div(\mathbf{v} \times \mathbf{c}), \tag{3.110}$$

for any fixed $\mathbf{c}$. Thus, with $grad(\mathbf{v} \times \mathbf{c}) = (\mathbf{v}_{,i} \times \mathbf{c}) \otimes \mathbf{g}^i$ we have

$$\mathbf{c} \cdot curl\mathbf{v} = \mathbf{g}^i \cdot \mathbf{v}_{,i} \times \mathbf{c} = \mathbf{c} \cdot \mathbf{g}^i \times \mathbf{v}_{,i}, \tag{3.111}$$

yielding

$$curl\mathbf{v} = \mathbf{g}^i \times \mathbf{v}_{,i} = v_{j;i} \mathbf{g}^i \times \mathbf{g}^j = \epsilon^{ijk} v_{j;i} \mathbf{g}_k. \tag{3.112}$$

Note that in this expression the covariant derivatives may be replaced by partial derivatives without affecting the result (why?).

We extend the curl operation to a tensor field $\mathbf{A}$ by requiring its curl, $curl\mathbf{A}$, to be the tensor field such that

$$(curl\mathbf{A})\mathbf{c} = curl(\mathbf{A}^t\mathbf{c}), \tag{3.113}$$

again for any fixed $\mathbf{c}$. Then,

$$(curl\mathbf{A})\mathbf{c} = \mathbf{g}^i \times (\mathbf{A}^t_{,i}\mathbf{c}). \tag{3.114}$$

From (3.81) we have

$$\mathbf{A}^t_{,i} = A^{jk}_{;i}\mathbf{g}_k \otimes \mathbf{g}_j \quad \text{and} \quad \mathbf{A}^t_{,i}\mathbf{c} = A^{jk}_{;i}\mathbf{g}_k(\mathbf{g}_j \cdot \mathbf{c}), \tag{3.115}$$

so that

$$\begin{aligned} \mathbf{g}^i \times (\mathbf{A}^t_{,i}\mathbf{c}) &= A^{jk}_{;i}\mathbf{g}^i \times \mathbf{g}_k(\mathbf{g}_j \cdot \mathbf{c}) = A^{jk}_{;i}g_{kl}\mathbf{g}^i \times \mathbf{g}^l(\mathbf{g}_j \cdot \mathbf{c}) \\ &= A^j_{\cdot l;i}\epsilon^{ilm}\mathbf{g}_m(\mathbf{g}_j \cdot \mathbf{c}) = (\epsilon^{ilm}A^j_{\cdot l;i}\mathbf{g}_m \otimes \mathbf{g}_j)\mathbf{c}, \end{aligned} \tag{3.116}$$

whence it follows, from (3.113), that

$$curl\mathbf{A} = \epsilon^{ilm}A^j_{\cdot l;i}\mathbf{g}_m \otimes \mathbf{g}_j. \tag{3.117}$$

Of central importance in later chapters is Stokes' theorem, stated here for Euclidean geometry. Thus, consider a simply connected patch of surface ω bounded by a smooth curve γ. Let ω be oriented by a unit-normal field $\mathbf{n}$, and let $\mathbf{y}(s)$ be the arclength parametrization of γ. We stipulate that γ be traversed in the direction $d\mathbf{y}$ such that $\boldsymbol{\nu}$, defined by $\boldsymbol{\nu} ds = d\mathbf{y} \times \mathbf{n}_{|\gamma}$, is the exterior unit normal to ω on γ. Then for any smooth vector field $\mathbf{v}(\mathbf{y})$, Stokes' theorem is the assertion that

$$\int_\gamma \mathbf{v} \cdot d\mathbf{y} = \int_\omega \mathbf{n} \cdot curl\mathbf{v}\, da. \tag{3.118}$$

This too can be extended to a smooth tensor field $\mathbf{A}(\mathbf{y})$ by taking $\mathbf{v} = \mathbf{A}^t\mathbf{c}$ with $\mathbf{c}$ an arbitrary fixed vector. Then,

$$\begin{aligned} \mathbf{c} \cdot \int_\gamma \mathbf{A}d\mathbf{y} &= \int_\gamma \mathbf{A}^t\mathbf{c} \cdot d\mathbf{y} = \int_\omega \mathbf{n} \cdot curl(\mathbf{A}^t\mathbf{c})da \\ &= \int_\omega \mathbf{n} \cdot (curl\mathbf{A})\mathbf{c}da = \mathbf{c} \cdot \int_\omega (curl\mathbf{A})^t\mathbf{n}da, \end{aligned} \tag{3.119}$$

yielding

$$\int_\gamma \mathbf{A}d\mathbf{y} = \int_\omega (curl\mathbf{A})^t\mathbf{n}da. \tag{3.120}$$

As a caution we note that this expression really has only a symbolic significance. In practice, to compute the integrals we replace the integrands by their component forms; in particular, by their Cartesian components. This restriction arises from the fact that the associated basis $\{\mathbf{i}_k\}$ is uniform and thus effectively "factors out" of the integrals. The resulting expressions have an invariant meaning, but this is lost if we replace the Cartesian components by those relative to a general system of coordinates. For we would not then recover an invariant statement on re-attaching basis elements. Stated differently, if the integrands are replaced by components relative to arbitrary coordinates, then the integrations entail the addition of vectors belonging to different tangent spaces, an operation having no invariant meaning. This observation also applies to the divergence theorem when applied to tensors, as stated in Chapter 1. This caution does not apply to (3.118) because the integrands therein are scalar fields, for which the notion of tangent space is irrelevant.

Consider now a parametrization $\xi^i(u^\alpha)$ of the surface ω in terms of a pair u^α; $\alpha = 1, 2$, of surface coordinates. These could be the longitudinal and latitudinal angles on the surface of a sphere, for example. Consider a pair of intersecting u^1- and u^2-coordinate curves passing through a point on the surface. On these curves we have $d\xi^i = d\zeta^i$ and $d\xi^i = d\eta^i$, respectively, where

$$d\zeta^i = \xi^i_{,1} du^1 \quad \text{and} \quad d\eta^i = \xi^i_{,2} du^2. \tag{3.121}$$

The oriented area measure of the surface is

$$\mathbf{n}da = \mathbf{g}_i d\zeta^i \times \mathbf{g}_j d\eta^j = \epsilon_{ijk} d\zeta^i d\eta^j \mathbf{g}^k, \tag{3.122}$$

where the basis elements are evaluated at the point in question. Using (3.112) we then have

$$curl\mathbf{v} \cdot \mathbf{n}da = \epsilon_{ijk} d\zeta^i d\eta^j \mathbf{g}^k \times \mathbf{g}^l \cdot \mathbf{v}_{,l} = \epsilon_{kij}\epsilon^{klm} d\zeta^i d\eta^j \mathbf{g}_m \cdot \mathbf{v}_{,l}. \tag{3.123}$$

The well-known $e - \delta$ identity satisfied by the permutation symbols (see Flügge's book, for example) may be used with (3.32) to conclude that

$$\epsilon_{kij}\epsilon^{klm} = \delta^l_i \delta^m_j - \delta^l_j \delta^m_i. \tag{3.124}$$

Then,

$$curl\mathbf{v} \cdot \mathbf{n}da = (d\zeta^l d\eta^m - d\zeta^m d\eta^l)\mathbf{g}_m \cdot \mathbf{v}_{,l}, \tag{3.125}$$

in which the parenthetical term, namely $(\xi^l_{,1}\xi^m_{,2} - \xi^m_{,1}\xi^l_{,2})du^1 du^2$, is the *wedge product* or *exterior product* $d\xi^l \wedge d\xi^m$. Then,

$$curl\mathbf{v} \cdot \mathbf{n}da = \mathbf{g}_m \cdot \mathbf{v}_{,l} d\xi^l \wedge d\xi^m = (v_{m,l} - v_k \Gamma^k_{ml}) d\xi^l \wedge d\xi^m = v_{m,l} d\xi^l \wedge d\xi^m, \tag{3.126}$$

where the final equality comes from $d\xi^l \wedge d\xi^m = -d\xi^m \wedge d\xi^l$, whereas $\Gamma^k_{ml} = \Gamma^k_{lm}$ in Euclidean space. Accordingly, (3.118) may be expressed in the simple form

$$\int_\gamma v_i d\xi^i = \int_\omega v_{m,l} d\xi^l \wedge d\xi^m. \tag{3.127}$$

Remarkably, this formula does not involve any geometry whatsoever, Euclidean or otherwise. It is, in fact, the fundamental statement of Stokes' theorem, valid for any smooth functions v_i defined on any manifold. Rather, the Euclidean form is properly interpreted as a special case, obtained by introducing the geometric quantities that appear in the course of reversing the steps in the derivation. The fundamental form is, however, more convenient for the considerations that follow.

To wit, consider the so-called *parallel transport* of a vector field $\mathbf{u}(t)$ defined on a curve with parametrization $\xi^i(t)$. This means that $\mathbf{u}$ is constant on the curve and hence that $\dot{\mathbf{u}}$, the derivative of $\mathbf{u}(t)$, vanishes. Writing $\mathbf{u} = u^i \mathbf{g}_i$ and invoking (3.61), we have

$$(\dot{u}^k + \Gamma^k_{ij} u^i \dot{\xi}^j)\mathbf{g}_k = \mathbf{0}, \tag{3.128}$$

implying that the parenthesis vanishes on the curve:

$$du^k = -\Gamma^k_{ij} u^i d\xi^j. \tag{3.129}$$

Granted initial conditions $u^k(t_0)$, the theory of ordinary differential equations guarantees the existence of unique functions $u^k(t)$ *on the curve*, but this by no means ensures that a vector *field* $u^k(\xi^i)$ exists in space such that $u^k(t) = u^k(\xi^i)$ on the curve $\xi^i(t)$. If such a field *did* exist, then we would have $du^k = u^k_{,j} d\xi^j$, and if parallel transport occurred on arbitrary curves, then the field would satisfy

$$u^k_{,j} = -\Gamma^k_{ij} u^i; \tag{3.130}$$

i.e., it would be covariantly constant. Further, if this field were twice differentiable, such that $u^k_{,jl} = u^k_{,lj}$, then it would be necessary that

$$(\Gamma^k_{ij} u^i)_{,l} = (\Gamma^k_{il} u^i)_{,j}. \tag{3.131}$$

Putting this another way, if the parallel field did exist, then the du^k would be exact differentials, and their closed-contour integrals would vanish:

$$0 = -\int_\gamma du^k = \int_\gamma \Gamma^k_{ij} u^i d\xi^j. \tag{3.132}$$

Taking the contour to bound a simply connected surface patch ω and applying Stokes' theorem, with $v_j = \Gamma^k_{ij} u^i$ and $k \in \{1, 2, 3\}$ fixed, we would then have

$$\begin{aligned} 0 &= \int_\omega (\Gamma^k_{ij} u^i)_{,l} d\xi^l \wedge d\xi^j = \int_\omega (\Gamma^k_{ij,l} u^i + \Gamma^k_{ij} u^i_{,l}) d\xi^l \wedge d\xi^j \\ &= \int_\omega (\Gamma^k_{ij,l} u^i - \Gamma^k_{ij} \Gamma^i_{ml} u^m) d\xi^l \wedge d\xi^j = \int_\omega (\Gamma^k_{mj,l} - \Gamma^k_{ij} \Gamma^i_{ml}) u^m d\xi^l \wedge d\xi^j \\ &= \frac{1}{2} \int_\omega R^k_{\cdot mlj} u^m d\xi^l \wedge d\xi^j, \end{aligned} \tag{3.133}$$

with

$$R^k_{\cdot mlj} = 2(\Gamma^k_{m[j,l]} - \Gamma^k_{i[j} \Gamma^i_{|m|l]}), \tag{3.134}$$

where we have observed the skew symmetry of the wedge product and adopted the convention that an index between vertical bars is exempted from skew symmetrization. For example,

$$A_{[j|m|l]} = \frac{1}{2}(A_{jml} - A_{lmj}), \tag{3.135}$$

so that

$$R^k_{\cdot mlj} = \Gamma^k_{mj,l} - \Gamma^k_{ml,j} + \Gamma^k_{il} \Gamma^i_{mj} - \Gamma^k_{ij} \Gamma^i_{ml}. \tag{3.136}$$

Thus, unless the $R^k_{\cdot mlj}$ vanish, a parallel *field* does not exist, whereas their vanishing ensures the existence of such a field in a neighborhood of the point in question. Section 3.7 of the book by Lovelock and Rund contains a particularly clear explanation of this phenomenon. Working in two dimensions, we find that parallel fields do not exist on the surface of a sphere (see Flügge's book). Because of examples like this, the $R^k_{\cdot mlj}$ are referred to as the components of the *curvature tensor*, although at this stage we have not shown that they actually satisfy the appropriate (fourth-order) tensor transformation formula. But we can derive some small comfort from the fact that

$$R^k_{\cdot m(lj)} = 0, \tag{3.137}$$

which of course is an immediate consequence of the definition (3.134).

Problem 3.17 Combine (3.130) and (3.131) to obtain $R^k_{\cdot mlj} u^m = 0$.

Problem 3.18 Show that (3.62) is true, with the gammas given by (3.56), if and only if $R^k_{\cdot mlj} = 0$.

Problem 3.19 Consider a vector field with covariant components v_i, and let $A_{ij} = v_{i;j}$. Suppose the v_i are twice differentiable, so that $v_{i,jk} = v_{i,kj}$.

(a) Show that, in a general affine space with both torsion and curvature,

$$A_{i[j;k]} + T^{l}_{\cdot jk} A_{il} = \frac{1}{2} R^{m}_{\cdot ijk} v_m.$$

(b) Note that the left-hand side of this equation is tensorial; i.e., it represents the covariant components of a third-order tensor. Use this fact to establish the transformation formula giving $\bar{R}^{m}_{\cdot ijk}$ in terms of the $R^{n}_{\cdot pqr}$ and hence conclude that $R^{m}_{\cdot ijk}$ are indeed the components of a fourth-order tensor. (You could proceed from (3.136) to reach the same conclusion, but the present method is considerably simpler.)

(c) It follows from the above formula that, unlike partial differentiation, covariant differentiation generally does not commute, i.e., $v_{i;jk} \neq v_{i;kj}$, where $v_{i;jk}$ is shorthand for $(v_{i;j})_{;k}$. Exceptionally, $v_{i;jk} = v_{i;kj}$ if the space is Euclidean ($T^{i}_{\cdot jk} = 0$ and $R^{m}_{\cdot ijk} = 0$). Prove this fact directly, using only the tensorial property of $v_{i;jk}$ and the fact that it is possible to parametrize Euclidean space in terms of Cartesian coordinates.

3.6 Torsion and the Weitzenböck connection

Consider now a space with connection coefficients of the form (see (3.56))

$$\Gamma^{k}_{ij} = \mathbf{g}^{k} \cdot \mathbf{g}_{i,j}. \tag{3.138}$$

Of course this subsumes Euclidean space, with the **g**'s constructed, per our recipe, from a position field $\mathbf{y}(\xi^{i})$. However, let us not confine ourselves to this case. Rather, let us consider this connection, called the *Weitzenböck connection*, on its own merits. That is, for the moment let us *not* require that the $\mathbf{g}_i$ be the derivatives of any position field. In this case the $\mathbf{g}_i$ are said to be *anholonomic*. In particular, this space may possess torsion because there is no a priori reason why the $\mathbf{g}_{[i,j]}$ should vanish.

To compute the curvature, we begin with

$$\Gamma^{k}_{mj,l} = \mathbf{g}^{k}_{,l} \cdot \mathbf{g}_{m,j} + \mathbf{g}^{k} \cdot \mathbf{g}_{m,jl}, \tag{3.139}$$

provided, of course, that the $\mathbf{g}'s$ are twice differentiable. We also have

$$\begin{aligned} \Gamma^{k}_{il}\Gamma^{i}_{mj} &= \Gamma^{k}_{sl}\Gamma^{i}_{mj}\delta^{s}_{i} = \Gamma^{k}_{sl}\Gamma^{i}_{mj}\mathbf{g}^{s} \cdot \mathbf{g}_{i} \\ &= \Gamma^{k}_{sl}\mathbf{g}^{s} \cdot \Gamma^{i}_{mj}\mathbf{g}_{i} = \Gamma^{k}_{sl}\mathbf{g}^{s} \cdot \mathbf{g}_{m,j} \\ &= -\mathbf{g}^{k}_{,l} \cdot \mathbf{g}_{m,j}. \end{aligned} \tag{3.140}$$

Then,

$$\Gamma^{k}_{mj,l} + \Gamma^{k}_{il}\Gamma^{i}_{mj} = \mathbf{g}^{k} \cdot \mathbf{g}_{m,jl}, \tag{3.141}$$

and (3.134) yields

$$R^{k}_{\cdot mlj} = 2\mathbf{g}^{k} \cdot \mathbf{g}_{m,[jl]}, \tag{3.142}$$

which, of course, vanishes identically because of the assumed degree of differentiability. Accordingly, any geometry with a Weitzenböck connection is flat. From this last formula we also have

$$\mathbf{g}_{m,[jl]} = (\mathbf{g}^{k} \cdot \mathbf{g}_{m,[jl]})\mathbf{g}_{k} = \frac{1}{2} R^{k}_{\cdot mlj}\mathbf{g}_{k}, \tag{3.143}$$

which implies that that the vanishing of the curvature components constitute the integrability conditions for the system (3.60), ensuring the existence of twice-differentiable fields $\mathbf{g}_i$ in a simply-connected region. This situation naturally subsumes Euclidean space, but, because torsion is allowed, flat spaces need not be Euclidean.

Problem 3.20 Assuming the differentials (3.60) to be exact, apply Stokes' theorem to arrive at an integral constraint involving the curvature tensor.

If the curvature does not vanish, then all we can say is that (3.60) is integrable on curves, but associated *fields* $\mathbf{g}_i$ do not exist in the space. In this case the geometry is not characterized by a connection of the form (3.138). An example is Riemannian space, characterized by a Levi-Civita connection (3.103) in which the metric is the basic variable. That is, whereas the Levi-Civita connection was derived from the Euclidean connection (3.56), it has an independent meaning as a connection in its own right. Taking it as the *definition* of the connection in terms of an arbitrary symmetric metric—tantamount to assuming a Riemannian geometry—and substituting into (3.136), we do *not* conclude that the curvature vanishes. Indeed gravity is essentially the same thing as curvature in the four-dimensional Riemannian space(-time) of general relativity (Schrödinger, 1950; Wald, 1984). In a general Riemannian geometry we naturally speak of the *Riemann curvature tensor.*

Before leaving this topic we note that we can define the "velocity" of a point on a curve with parametrization $\xi^i(t)$ by $\mathbf{g}_i\dot{\xi}^i$. We denote this by $\dot{\mathbf{y}}$, i.e.,

$$d\mathbf{y} = \mathbf{g}_i d\xi^i. \tag{3.144}$$

Naturally we are curious to know whether this is integrable, that is, whether an associated position *field* $\mathbf{y}(\xi^i)$ exists. From all that has been said, we would expect this to be answerable in the affirmative if the connection were of Weitzenböck type with zero torsion, for we would then be back in Euclidean space. To confirm this we apply Stokes' theorem, keeping in mind the skew symmetry of the wedge product:

$$\int_{\gamma} d\mathbf{y} = \int_{\gamma} \mathbf{g}_i d\xi^i = \int_{\omega} \mathbf{g}_{i,j} d\xi^j \wedge d\xi^i = \int_{\omega} \Gamma^{k}_{[ij]}\mathbf{g}_k d\xi^j \wedge d\xi^i. \tag{3.145}$$

Therefore, $d\mathbf{y}$ is not exact, and a position field does not exist, if the connection (3.138) possesses torsion. If the torsion vanishes, then $d\mathbf{y}$ *is* exact, and $\mathbf{g}_i = \mathbf{y}_{,i}$ (see (3.2)).

We will see that this scenario describes a body containing a continuous distribution of dislocations, the evolution of which in space and time is intimately connected to plasticity in crystalline materials. In this context the left-most integral is the *Burgers* vector of the distribution, and the torsion may then be interpreted as a dislocation density. This observation effectively inaugurated the geometric school of material *defects*. The seminal papers by Bilby et al., Eckart, Kondo, Kröner, Noll, and Wang, and the monograph by Epstein, are essential reading for anyone interested in learning more about the foundations of this beautiful subject. A comprehensive survey, written at the level of this book, is given in the review article by Gairola.

References

Bilby, B. A., Bullough, R., and Smith, E. (1955) Continuous distributions of dislocations: A new application of the methods of non-Riemannian geometry. *Proc. Roy. Soc. Lond.* A231, 263–73.

Brillouin, L. (1964). *Tensors in Mechanics and Elasticity.* Academic Press, New York.

Cartan, E. (2001). *Riemannian Geometry in an Orthogonal Frame* (transl. by V. V. Goldberg). World Scientific, Singapore.

Ciarlet, P. G. (2005). *An Introduction to Differential Geometry with Applications to Elasticity.* Springer, Dordrecht.

Eckart, C. (1948). The thermodynamics of irreversible processes. IV: The theory of elasticity and anelasticity. *Phys. Rev.* 73, 373–82.

Epstein, M. (2010). *The Geometrical Language of Continuum Mechanics.* Cambridge University Press, Cambridge, UK.

Flügge, W. (1972) *Tensor Analysis and Continuum Mechanics.* Springer, Berlin.

Gairola, B. K. D. (1979). Non-linear elastic problems. In: Nabarro, F. R. N. (Ed.), *Dislocations in Solids,* Vol. 1. North-Holland, Amsterdam.

Green, A. E., and Zerna, W. (1968). *Theoretical Elasticity.* Oxford University Press, Oxford.

Kondo, K. (1955) Non-Riemannian geometry of imperfect crystals from a macroscopic viewpoint. In: Kondo, K. (Ed.), *RAAG Memoirs of the Unifying Study of Basic Problems in Engineering Sciences by Means of Geometry,* Vol. 1. Gakujutsu Bunken Fukyu-Kai, Tokyo.

Kröner, E. (1960) Allgemeine kontinuumstheorie der versetzungen und eigenspannungen. *Arch. Ration. Mech. Anal.* 4, 273–334.

Lichnerowicz, A. (1962). *Tensor Calculus.* Methuen, London.

Lovelock, D., and Rund, H. (1989). *Tensors, Differential Forms and Variational Principles.* Dover, New York.

Noll, W. (1967). Materially uniform simple bodies with inhomogeneities. *Arch. Ration. Mech. Anal.* 27, 1–32.

Povstenko, Y. Z. (1991). Connection between non-metric differential geometry and the mathematical theory of imperfections. *Int. J. Engng. Sci.* 29, 37–46.

Schouten, J. A. (1954). *Ricci-Calculus.* Springer, Berlin.

Schrödinger, E. (1950). *Space-Time Structure.* Cambridge University Press, Cambridge, UK.

Simmonds, J. G. (1994). *A Brief on Tensor Analysis.* Springer, New York.

Sokolnikoff, I. S. (1964). *Tensor Analysis: Theory and Applications to Geometry and Mechanics of Continua.* Wiley, New York.

Stoker, J. J. (1989). *Differential Geometry.* Wiley-Interscience, New York.

Szerkeres, P. (2004). *A Course in Modern Mathematical Physics.* Cambridge University Press, Cambridge, UK.

Wald, R. M. (1984). *General Relativity.* University of Chicago Press, Chicago.

Wang, C.-C. (1967). On the geometric structure of simple bodies, a mathematical foundation for the theory of continuous distributions of dislocations. *Arch. Ration. Mech. Anal.* 27, 33–94.

Willmore, T. J. (1959). *Differential Geometry.* Oxford University Press, Oxford.

4

Deformation and stress in convected coordinates

Having surveyed the rudiments of tensor analysis and differential geometry in curvilinear coordinates, it is appropriate and instructive to apply what we've learned to the analysis of deformation and stress.

4.1 Deformation gradient and strain

To this end we will use the notation of the previous chapter to describe the geometry of the Euclidean space in which the current configuration, κ_t, of a body resides. To distinguish this from the Euclidean geometry of the reference configuration κ, we adopt a different notation. Thus, in κ we denote the position field by $\mathbf{x}$ rather than $\mathbf{y}$, as in Chapter 1, and take it to be a function $\mathbf{x}(\xi^i)$, where $\{\xi^i\}$ is a set of right-handed curvilinear coordinates. Inserting this into the deformation function (1.9), we have a parametrization of κ_t in terms of the same coordinates:

$$\mathbf{y}(\xi^i, t) = \boldsymbol{\chi}(\mathbf{x}(\xi^i), t). \tag{4.1}$$

We can now bring all the apparatus of Chapter 3 to bear on this position field at each fixed time t. By using a single coordinate system ξ^i to parametrize both κ and κ_t, we are in effect taking the coordinates to convect with the deformation of the body. That is, as in Chapter 3 we suppose the coordinates to be in one-to-one correspondence with the position $\mathbf{x}$, at least locally, and hence in one-to-one correspondence with a material point of the body. In this way each material point is identified with a unique triplet $\{\xi^i\}$, the elements of which then serve as its permanent labels. In particular, the material derivatives $\dot{\xi}^i$ vanish, and the material velocity—the time derivative of $\mathbf{y}$ holding $\mathbf{x}$ fixed—is simply the partial time derivative in the convected coordinate formalism. A good way to visualize this situation is to imagine drawing a curvilinear coordinate grid on a balloon. Suppose a particular material point on the balloon has the coordinates $(1, 2)$ in this (two-dimensional) grid. As the balloon is inflated the grid deforms with it, and the label $(1, 2)$ continues to identify the same material point. The coordinate curves are thus embedded

A Course on Plasticity Theory. David J. Steigmann, Oxford University Press. © David J. Steigmann (2022).
DOI: 10.1093/oso/9780192883155.003.0004

in the material; that is, they are material curves. The use of ξ^i and t as independent variables is therefore tantamount to a material, or Lagrangian, description of the motion.

Convected coordinates are not as widely used in continuum mechanics as they once were. This is unfortunate, because they furnish a particularly useful tool for analysts who have made the modest investment of effort, as we have, needed to learn the prerequisite tensor analysis. They are an indispensable tool in the study of rods and shells, for example, regarded as one- and two-dimensional continua respectively. The books by Sedov, Green and Zerna, and Green and Adkins are excellent sources of information on the practical use of convected coordinates.

Using $\mathbf{x}(\xi^i)$ we define the elements

$$\mathbf{e}_i = \mathbf{x}_{,i} \tag{4.2}$$

of the referential tangent basis, and their reciprocals

$$\mathbf{e}^i = \nabla\xi^i, \tag{4.3}$$

where ∇ is the gradient with respect to $\mathbf{x}$, i.e., $\mathbf{e}^i = \nabla\hat{\xi}^i(\mathbf{x})$. We then compute the associated metric and reciprocal metric components

$$e_{ij} = \mathbf{e}_i \cdot \mathbf{e}_j \quad \text{and} \quad e^{ij} = \mathbf{e}^i \cdot \mathbf{e}^j, \tag{4.4}$$

and the connection coefficients

$$\bar{\Gamma}^m_{ij} = \mathbf{e}^m \cdot \mathbf{e}_{i,j} = \tfrac{1}{2}e^{mk}(e_{ki,j} + e_{kj,i} - e_{ij,k}), \tag{4.5}$$

etc., together with the counterparts of the various concepts discussed in Chapter 3. Here, however, we use an overbar on the gammas and a vertical bar to denote the associated covariant derivatives. For example, the gradient of $\mathbf{v} = v^i\mathbf{e}_i$ with respect to $\mathbf{x}$ is

$$\nabla\mathbf{v} = v^k_{|i}\mathbf{e}_k \otimes \mathbf{e}^i, \tag{4.6}$$

where

$$v^k_{|i} = v^k_{,i} + \bar{\Gamma}^k_{ji}v^j. \tag{4.7}$$

A convected-coordinate representation of the deformation gradient $\mathbf{F}$ defined in Chapter 1 is easily derived. Fixing the time t, from (3.2) and (4.3) we have

$$\mathbf{F}d\mathbf{x} = d\mathbf{y} = \mathbf{g}_i d\xi^i = \mathbf{g}_i(\mathbf{e}^i \cdot d\mathbf{x}) = (\mathbf{g}_i \otimes \mathbf{e}^i)d\mathbf{x}, \tag{4.8}$$

and as this purports to hold for arbitrary $d\mathbf{x}$, it follows that

$$\mathbf{F} = \mathbf{g}_i \otimes \mathbf{e}^i. \tag{4.9}$$

This maps vectors in T_κ, the tangent space to κ, to vectors in T_{κ_t}, the tangent space to κ_t, and is accordingly referred to as a *two-point tensor*. Using $\mathbf{F}^t = \mathbf{e}^i \otimes \mathbf{g}_i$, the right

Cauchy–Green deformation tensor is then given by

$$\mathbf{C} = (\mathbf{F}^t\mathbf{g}_i) \otimes \mathbf{e}^i = g_{ij}\mathbf{e}^i \otimes \mathbf{e}^j. \tag{4.10}$$

The covariant components of this tensor are thus none other than the components of the metric induced by the coordinates in κ_t.

Problem 4.1 The contravariant components of $\mathbf{C}$ are *not* equal to g^{ij}. What are they?

Problem 4.2 Show that e^{ij} are the contravariant components of the left Cauchy–Green deformation tensor $\mathbf{B} = \mathbf{F}\mathbf{F}^t$.

It is conceptually helpful to distinguish the Euclidean tangent spaces T_{κ_t} and T_κ. Thus, we identify the unit tensor **i**, introduced in Chapter 3, as the *spatial* unit tensor that maps any vector in T_{κ_t} to itself. In the same way, pursuant to the discussion in Chapter 1, let **I** be the *referential* unit tensor, mapping any vector in T_κ to itself. Proceeding as in (3.47), we have

$$\mathbf{I} = \mathbf{e}_i \otimes \mathbf{e}^i = e_{ij}\mathbf{e}^i \otimes \mathbf{e}^j, \tag{4.11}$$

together with the counterparts of the various other representations found there. From (1.20) and (4.10) we then derive the Lagrange strain

$$\mathbf{E} = E_{ij}\,\mathbf{e}^i \otimes \mathbf{e}^j, \tag{4.12}$$

where

$$E_{ij} = \tfrac{1}{2}(g_{ij} - e_{ij}). \tag{4.13}$$

This rather elegant formula indicates that the covariant components of the strain are just one-half the difference of the metrics in κ_t and κ at the material point with coordinates ξ^i. However, the contravariant components are *not* proportional to the difference of the reciprocal metrics. Be aware that this situation is typical of the convected-coordinate formalism: Results that hold for components of one kind—covariant, contravariant, or mixed—are not valid for other kinds.

The determinant $\mathcal{J}_F$ of the deformation gradient is given by the ratio

$$\mathcal{J}_F = \mathbf{F}\mathbf{e}_1 \cdot \mathbf{F}\mathbf{e}_2 \times \mathbf{F}\mathbf{e}_3/\mathbf{e}_1 \cdot \mathbf{e}_2 \times \mathbf{e}_3 \tag{4.14}$$

of scalar triple products. See the discussion in Chadwick excellent pocketbook, for example. Now (4.9) implies that the numerator is simply $\mathbf{g}_1 \cdot \mathbf{g}_2 \times \mathbf{g}_3$. In Chapter 3 we showed that this is given by $\sqrt{g}$, where $g = \det(g_{ij})$. In the same way, $\mathbf{e}_1 \cdot \mathbf{e}_2 \times \mathbf{e}_3 = \sqrt{e}$, where

$$e = \det(e_{ij}). \tag{4.15}$$

Then,

$$\mathcal{J}_F = \sqrt{g/e}. \tag{4.16}$$

Thus, the relation (1.22) between the referential and spatial mass densities may be expressed, in terms of convected coordinates, as

$$\rho_\kappa \sqrt{e} = \rho \sqrt{g}. \tag{4.17}$$

In Chapter 3 we also discussed the oriented surface area measure given by (3.122), i.e., $\mathbf{n}da = \epsilon_{ijk} d\zeta^i d\eta^j \mathbf{g}^k$, where $\epsilon_{ijk} = \sqrt{g} e_{ijk}$. We interpret this as pertaining to a surface $\omega \subset \kappa_t$. Suppose this to be a material surface, so that its image $\Omega \subset \kappa$ consists of the same material points. Recalling that the coordinates are convected with these points, we conclude that the area measure on Ω is

$$\boldsymbol{\nu} dA = \bar{\epsilon}_{ijk} d\zeta^i d\eta^j \mathbf{e}^k, \quad \text{where} \quad \bar{\epsilon}_{ijk} = \sqrt{e} e_{ijk}, \tag{4.18}$$

and $\boldsymbol{\nu}$ is a unit-normal field. Accordingly, the *covariant* components of these area measures, relative to their respective reciprocal bases, are related by

$$n_k da = \mathcal{J}_F \nu_k dA. \tag{4.19}$$

Reattaching the relevant bases, we arrive at the Piola–Nanson formula (1.35),

$$\mathbf{n}da = \mathcal{J}_F \nu_k \mathbf{g}^k dA = \mathcal{J}_F \mathbf{g}^k (\mathbf{e}_k \cdot \boldsymbol{\nu}) dA = \mathbf{F}^* \boldsymbol{\nu} dA, \tag{4.20}$$

where

$$\mathbf{F}^* = \mathcal{J}_F \mathbf{g}^k \otimes \mathbf{e}_k \tag{4.21}$$

is the cofactor of the deformation gradient.

Problem 4.3 Show that $\mathbf{F}^{-t} = \mathbf{g}^k \otimes \mathbf{e}_k$ and hence that (4.21) is consistent with (1.36).

4.2 Strain compatibility

Later, in Chapter 5, we will devote considerable space to a discussion of strain compatibility. This entails a set of restrictions on the strain components arising from the premise—a prejudice rooted in our terrestrial experience—that bodies always occupy regions of Euclidean space as they deform. The connections $\Gamma_{ij}^k = \mathbf{g}^k \cdot \mathbf{g}_{i,j}$ and $\bar{\Gamma}_{ij}^k = \mathbf{e}^k \cdot \mathbf{e}_{i,j}$ are therefore torsionless and curvature-free. The strain compatibility conditions may then be obtained by solving (4.13) for the metric g_{ij} in terms of the strain E_{ij}, substituting the result into the Levi-Civita formula (3.103) for Γ_{ij}^k, and requiring that the Riemannian curvature components (3.136) vanish, together with their referential counterparts $\bar{R}^k_{\cdot mlj}$.

Before undertaking this arduous task it is worthwhile to first look for possible symmetries of the curvature arising from the Levi-Civita connection. To this end it is convenient to introduce the fully covariant curvature

$$R_{pmlj} = g_{pk} R^k_{\cdot mlj}. \tag{4.22}$$

Then from (3.137) we have

$$R_{pm(lj)} = 0, \tag{4.23}$$

whether or not the connection is Levi-Civita. We intend to show that if it *is* Levi-Civita, then the *major symmetry*

$$R_{pmlj} = R_{ljpm} \tag{4.24}$$

obtains, so that, in addition,

$$R_{(pm)lj} = 0. \tag{4.25}$$

Accordingly, in this case we have the significant simplification

$$R_{pmlj} = R_{[pm][lj]}. \tag{4.26}$$

The demonstration of this important result is somewhat tedious. We'll endeavor to make it as brief and straightfoward as possible. To begin, we combine (3.136) and (4.22) to obtain

$$R_{pmlj} = g_{pk}(\Gamma^k_{mj,l} + \Gamma^k_{il}\,\Gamma^i_{mj}) - g_{pk}(\Gamma^k_{ml,j} + \Gamma^k_{ij}\,\Gamma^i_{ml}), \tag{4.27}$$

in which (3.97) is used to write

$$g_{pk}\Gamma^k_{mj,l} = (g_{pk}\Gamma^k_{mj})_{,l} - g_{pk,l}\Gamma^k_{mj} = \Gamma_{pmj,l} - g_{pk,l}\Gamma^k_{mj}. \tag{4.28}$$

On combining this with (3.98), the first set of parentheses in (4.27) becomes

$$g_{pk}(\Gamma^k_{mj,l} + \Gamma^k_{il}\,\Gamma^i_{mj}) = \Gamma_{pmj,l} - \Gamma^i_{mj}\,\Gamma_{ipl}, \tag{4.29}$$

and application of a similar reduction to the second set of parentheses then yields

$$R_{pmlj} = \Gamma_{pmj,l} - \Gamma_{ljp,m} + \Gamma^i_{ml}\,\Gamma_{ipj} - \Gamma^i_{mj}\,\Gamma_{ipl}. \tag{4.30}$$

We can now form the difference of the left- and right-hand sides of (4.24). Noting that $\Gamma^i_{jp}\Gamma_{ilm} = \Gamma_{ijp}\Gamma^i_{lm} = \Gamma^i_{ml}\Gamma_{ipj}$ and that $\Gamma^i_{jm}\Gamma_{ilp} = \Gamma^i_{mj}\Gamma_{ipl}$, we find that in this difference the products of the gammas cancel out, leaving only derivative terms

$$R_{pmlj} - R_{ljpm} = \Gamma_{pmj,l} + \Gamma_{ljp,m} - \left(\Gamma_{pml,j} + \Gamma_{ljm,p}\right). \tag{4.31}$$

Substituting (3.101), we then have

$$\begin{aligned} \tfrac{1}{2}\left(R_{pmlj} - R_{ljpm}\right) &= \left(g_{pm,jl} - g_{pm,lj}\right) + \left(g_{lj,pm} - g_{lj,mp}\right) + \left(g_{jm,lp} - g_{mj,pl}\right) \\ &\quad + \left(g_{jp,ml} - g_{jp,lm}\right) + \left(g_{ml,pj} - g_{ml,jp}\right) + \left(g_{pl,jm} - g_{lp,mj}\right), \end{aligned} \tag{4.32}$$

in which each set of parentheses vanishes separately, provided, as we assume, that the metric is twice differentiable. Thus, the symmetry condition (4.24) holds in Riemannian—hence also in Euclidean—space, wherein the connection is Levi-Civita.

This major simplification leads to (4.26), and hence to the conclusion that the fourth-order Riemann tensor stands in one-to-one relation to the second-order *Einstein tensor* with contravariant components defined by

$$\Pi^{ij} = \tfrac{1}{4}\epsilon^{ikl}\epsilon^{jmn}R_{klmn}. \tag{4.33}$$

To prove this claim we need only invert this relation to obtain

$$R_{pqrs} = \epsilon_{ipq}\epsilon_{jrs}\Pi^{ij}. \tag{4.34}$$

Observe that (4.26) is indeed satisfied.

Problem 4.4 Show that (4.33) implies (4.34). Hint: Make repeated use of (3.124).

Accordingly the components of the Riemann curvature vanish if and only if the Π^{ij} vanish. Beyond this, from (4.24) and (4.33), and with a relabeling of summation indices, we have

$$\begin{aligned} \Pi^{ji} &= \tfrac{1}{4}\epsilon^{jkl}\epsilon^{imn}R_{klmn} = \tfrac{1}{4}\epsilon^{jkl}\epsilon^{imn}R_{mnkl} \\ &= \tfrac{1}{4}\epsilon^{jmn}\epsilon^{ikl}R_{klmn} = \Pi^{ij}. \end{aligned} \tag{4.35}$$

Thus, the three independent components $\Pi^{[ij]}$ vanish identically, leaving six non-trivial compatibility conditions. We can put these into another, arguably more convenient, form in terms of the *Ricci tensor*, the second-order tensor with covariant components defined by

$$R_{ij} = R^k_{\cdot jik}. \tag{4.36}$$

The relationship between the Einstein and Ricci tensors may be established with the aid of the identity (see Eq. 3.19 in Flügge's book)

$$\epsilon^{ijk}\epsilon^{lmn} = g^{il}(g^{jm}g^{kn} - g^{jn}g^{km}) + g^{in}(g^{jl}g^{km} - g^{jm}g^{kl}) + g^{im}(g^{jn}g^{kl} - g^{jl}g^{kn}), \quad (4.37)$$

yielding

$$4\Pi^{il} = \epsilon^{ijk}\epsilon^{lmn}R_{jkmn} = g^{il}(R^{mn}_{\cdot\cdot mn} - R^{nm}_{\cdot\cdot mn}) + g^{in}(R^{lm}_{\cdot\cdot mn} - R^{ml}_{\cdot\cdot mn}) + g^{im}(R^{nl}_{\cdot\cdot mn} - R^{ln}_{\cdot\cdot mn}), \quad (4.38)$$

where $R^{ij}_{\cdot\cdot kl} = g^{jm}R^{i}_{\cdot mkl}$. Using $R^{ij}_{\cdot\cdot kl} = -R^{ji}_{\cdot\cdot kl}$, which follows from (4.26), we reduce this to

$$2\Pi^{il} = g^{in}R^{lm}_{\cdot\cdot mn} + g^{im}R^{nl}_{\cdot\cdot mn} - Rg^{il}, \quad (4.39)$$

where

$$R = R^{nm}_{\cdot\cdot mn} = R^{n}_{\cdot n}, \quad (4.40)$$

in which $R^{i}_{\cdot j} = g^{ik}R_{kj}$, is called the *scalar curvature*. Raising indices in the usual way, we use $R^{ij} = R^{kji}_{\cdot\cdot\cdot k}$ with $g^{im}R^{nl}_{\cdot\cdot mn} = R^{nli}_{\cdot\cdot\cdot n}(= R^{il})$ and $g^{in}R^{lm}_{\cdot\cdot mn} = R^{lm\cdot i}_{\cdot\cdot m} = R^{mli}_{\cdot\cdot\cdot m}(= R^{il})$, which follows from (4.26), finally reaching

$$\Pi^{il} = R^{il} - \tfrac{1}{2}Rg^{il}. \quad (4.41)$$

It is customary to take this as the definition of the Einstein tensor in n dimensions; for example, in relativity theory ($n = 4$), whereas the formula (4.33) is valid in three dimensions only.

Problem 4.5 Invert this relation to obtain $R^{ij} = \Pi^{ij} - \Pi g^{ij}$, where $\Pi = \Pi^{n}_{\cdot n}$ and $\Pi^{i}_{\cdot j} = g_{jk}\Pi^{ik}$.

Problem 4.6 Show that the Einstein tensor is divergence-free, i.e., $\Pi^{ij}_{\cdot ;j} = 0$. This is a consequence of the Bianchi identities (see, for example, Section 86 of Lichnerowicz' book, cited in Chapter 3). Incidentally, this result plays a decisive role in Einstein's theory—hence the name.

Thus, the relation between the Einstein and Ricci tensors is one-to-one. In three dimensions the Riemann, Einstein, and Ricci tensors are therefore equivalent. Note that we have established the symmetry of the Ricci tensor in three dimensions, so that the six strain compatibility conditions may be expressed simply as

$$R_{ij} = 0. \quad (4.42)$$

Symmetry in n dimensions is typically proved by using (3.136) and (4.36) to write

$$R_{ij} = \Gamma^k_{jk,i} - \Gamma^k_{ji,k} + \Gamma^k_{mi}\Gamma^m_{jk} - \Gamma^k_{mk}\Gamma^m_{ji}, \tag{4.43}$$

in which the i, j -symmetry of the second and fourth terms on the right-hand side is obvious. The symmetry of the third term follows from $\Gamma^k_{mi}\Gamma^m_{jk} = \Gamma^m_{ki}\Gamma^k_{jm} = \Gamma^k_{mj}\Gamma^m_{ik}$, and that of the first term from $\Gamma^k_{jk,i} = (\ln\sqrt{g})_{,ji}$, which in turn follows from (3.107).

For the purpose of exhibiting the role of the strain in (4.42), it is convenient to express the components R_{ij} in terms of covariant derivatives with respect to the referential connection $\bar{\Gamma}^k_{ij}$. To this end we introduce

$$S^k_{\cdot ij} = \Gamma^k_{ij} - \bar{\Gamma}^k_{ij}. \tag{4.44}$$

This is the difference of two connections induced by a given system of (convected) coordinates. Having solved Problem 3.16(a), we know that these are the components of a third-order tensor.

Problem 4.7 Prove this.

Problem 4.8 Demonstrate that the symmetry of the Ricci tensor follows directly from (4.23)–(4.25).

Equation (4.43) then furnishes

$$\begin{aligned} R_{ij} &= S^k_{\cdot jk,i} - S^k_{\cdot ji,k} + S^k_{\cdot mi}S^m_{\cdot jk} - S^k_{\cdot mk}S^m_{\cdot ji} \\ &\quad + S^k_{\cdot mi}\bar{\Gamma}^m_{jk} - S^k_{\cdot mk}\bar{\Gamma}^m_{ji} + S^m_{\cdot jk}\bar{\Gamma}^k_{mi} - S^m_{\cdot ji}\bar{\Gamma}^k_{mk} + \bar{R}_{ij}, \end{aligned} \tag{4.45}$$

where $\bar{R}_{ij}$—the components of the referential Ricci tensor—are obtained by substituting $\bar{\Gamma}^k_{ij}$ in place of Γ^k_{ij}. They vanish, of course, because the geometry of κ is Euclidean, but the equation remaining is still rather formidable. To simplify it we use an old geometer's trick: We choose the coordinates ξ^i to be Cartesian in κ, so that the $\bar{\Gamma}^k_{ij}$ vanish, and with them the second line in (4.45). The first two terms, involving partial derivatives of the S's, are then the Cartesian representations of the corresponding referential covariant derivatives. Because the first line involves tensors exclusively, we can then use the tensor transformation rules established in Chapter 3 to conclude that

$$R_{ij} = S^k_{\cdot jk|i} - S^k_{\cdot ji|k} + S^k_{\cdot mi}S^m_{\cdot kj} - S^k_{\cdot mk}S^m_{\cdot ji} \tag{4.46}$$

in arbitrary coordinates.

To achieve our aim we must represent the $S^k_{\cdot ij}$ in terms of the strain components. Using the geometer's trick, we can again select Cartesian coordinates in κ to reduce (4.44) to

$S^{k}_{\cdot ij} = \Gamma^{k}_{ij}$, with Γ^{k}_{ij} given by (3.103). Reverting to general coordinates and invoking Ricci's lemma in the form $e_{ij|k} = 0$, we obtain

$$S^{k}_{\cdot ij} = g^{kl}(E_{li|j} + E_{lj|i} - E_{ij|l}). \tag{4.47}$$

Problem 4.9 It is not immediately obvious that the right-hand side of (4.46) possesses the required symmetry with respect to interchange of i and j. The symmetry of the second and fourth terms *is* obvious, while that of the third follows from $S^{k}_{\cdot mi}S^{m}_{\cdot kj} = S^{m}_{\cdot ki}S^{k}_{\cdot mj} = S^{k}_{\cdot mj}S^{m}_{\cdot ki}$. As for the first term, establish that

$$S^{k}_{\cdot jk} = (\mathcal{J}_F)_{,j}/\mathcal{J}_F,$$

and use this to show that $S^{k}_{\cdot jk|i} = S^{k}_{\cdot ik|j}$.

Problem 4.10 Obtain the strain compatibility conditions by substituting (4.47) into (4.46). Along the way you will encounter terms like $g^{jm}_{|i}$. Show that $g^{jm}_{|i} = -g^{jk}g^{ml}g_{kl|i} = -2g^{jk}g^{ml}E_{kl|i}$ and use this in $R_{ij} = 0$ to derive the long-awaited strain compatibility equations. These indicate that the strain field cannot be arbitrary, but instead that it satisfies certain differential constraints.

4.3 Stress, equations of motion

Moving on to stress, recall that the Cauchy stress $\mathbf{T}$ maps spatial vectors—the orientation $\mathbf{n}$ of a surface in κ_t, for example—to other spatial vectors; for example, the traction $\mathbf{t}$ acting on the tangent plane to the surface. Accordingly we decompose $\mathbf{T}$ in bases constructed from the $\mathbf{g}_i$. One such decomposition is

$$\mathbf{T} = T^{ij}\mathbf{g}_i \otimes \mathbf{g}_j, \tag{4.48}$$

in terms of contravariant components. We then find that (1.32) and (1.34) are equivalent to

$$T^{ij}_{;j} + \rho b^i = \rho a^i \quad \text{and} \quad T^{ij} = T^{ji}, \tag{4.49}$$

respectively, where $b^i = \mathbf{b} \cdot \mathbf{g}^i$ are the body-force components, $T^{ij}_{;j}$ is the covariant divergence based on the Levi-Civita connection (3.103), and of course $a^i = \dot{\mathbf{v}} \cdot \mathbf{g}^i$ are the components of the acceleration in which $\mathbf{v} = \dot{\mathbf{y}}$ is the material velocity. In particular,

$$\begin{aligned} a^i &= \mathbf{g}^i \cdot (v^j\mathbf{g}_j)^{\cdot} = \mathbf{g}^i \cdot (\dot{v}^j\mathbf{g}_j + v^j\dot{\mathbf{y}}_{,j}) \\ &= \mathbf{g}^i \cdot (\dot{v}^j\mathbf{g}_j + v^j\mathbf{v}_{,j}) = \dot{v}^i + v^i_{;j}v^j, \end{aligned} \tag{4.50}$$

where we have used the independence of ξ^i and t to commute the coordinate and time derivatives $(\dot{\mathbf{y}})_{,j} = (\mathbf{y}_{,j})^{\cdot}$, this being permissible if the velocity is twice differentiable with respect to the coordinates and time jointly. The last of these equalities is the well-known formula for the acceleration in terms of the *convective* derivative of the velocity. This is often a difficult concept for the beginning student, whereas the result is immediate in the convected-coordinate framework.

Combining (4.21) and (4.48), we find that the Piola stress **P**, defined by (1.38), admits the representation

$$\mathbf{P} = \mathbf{P}^i \otimes \mathbf{e}_i, \quad \text{where} \quad \mathbf{P}^i = J_F T^{ji}\mathbf{g}_j. \tag{4.51}$$

This, like the deformation gradient, is a two-point tensor. Therefore the Piola–Kirchhoff stress **S**, defined in (1.42), has the representation

$$\mathbf{S} = S^{ij}\mathbf{e}_i \otimes \mathbf{e}_j, \quad \text{where} \quad S^{ij} = J_F T^{ij}, \tag{4.52}$$

the latter holding for contravariant components only. Further, the symmetry condition $(4.49)_2$—equivalent to $T^{ij}\mathbf{g}_i \times \mathbf{g}_j = \mathbf{0}$ —imposes the restriction

$$\mathbf{g}_i \times \mathbf{P}^i = \mathbf{0} \tag{4.53}$$

on the *Piola stress vectors* $\mathbf{P}^i$.

Problem 4.11 Prove that $S^{ij} = J_F T^{ij}$, as claimed.

Problem 4.12 Derive the Doyle–Ericksen formula

$$J_F T^{ij} = 2\frac{\partial W}{\partial g_{ij}}$$

for hyperelastic materials, where W is the strain-energy function.

Problem 4.13 Show that $\mathbf{P}^i = \mathbf{P}\mathbf{e}^i$ and hence that $\mathbf{P}^i$ is proportional to the traction exerted on a surface where ξ^i, with $i \in \{1, 2, 3\}$ fixed, is constant.

Using the referential counterparts of (3.80) and (3.107), the divergence of the Piola stress occurring in (1.40) is found to be

$$\begin{aligned} Div\mathbf{P} &= (\mathbf{P})_{,j}\mathbf{e}^j = (\mathbf{P}^i \otimes \mathbf{e}_i)_{,j}\mathbf{e}^j = (\mathbf{P}^i_{,j} \otimes \mathbf{e}_i + \mathbf{P}^i \otimes \mathbf{e}_{i,j})\mathbf{e}^j \\ &= \mathbf{P}^i_{,i} + \bar{\Gamma}^j_{ij}\mathbf{P}^i = \tfrac{1}{\sqrt{e}}(\sqrt{e}\mathbf{P}^i)_{,i}. \end{aligned} \tag{4.54}$$

In the same way, from

$$\mathbf{T} = \mathbf{T}^i \otimes \mathbf{g}_i, \quad \text{where} \quad \mathbf{T}^i = T^{ji}\mathbf{g}_j \tag{4.55}$$

are the Cauchy stress vectors, we have

$$divT = \frac{1}{\sqrt{g}}(\sqrt{g}\mathbf{T}^i)_{,i}. \tag{4.56}$$

Substituting this into (1.32) and multiplying the result by $\mathcal{J}_F$, with the aid of (4.16) and $(4.51)_2$ we immediately arrive at (1.40): The convected-coordinate formalism facilitates a trivial derivation of the Piola equations of motion from the Cauchy equations, and vice versa. The conventional derivation—see Chadwick's book, for example—is based on the formula

$$Div(\mathbf{TF}^*) = \mathcal{J}_F div\mathbf{T}, \tag{4.57}$$

which in turn is a consequence of the Piola identity $Div\mathbf{F}^* = \mathbf{0}$.

Problem 4.14 Prove this assertion.

It's particularly easy to establish the Piola identity by using convected coordinates. Thus, we combine (4.16), (4.21), $(4.51)_1$ and the formula (4.54) for the divergence, with $\mathbf{P}^i$ replaced by $\mathcal{J}_F\mathbf{g}^i$:

$$\begin{aligned} Div\mathbf{F}^* &= \frac{1}{\sqrt{e}}(\sqrt{e}\mathcal{J}_F\mathbf{g}^i)_{,i} = \frac{\mathcal{J}_F}{\sqrt{g}}(\sqrt{g}\mathbf{g}^i)_{,i} \\ &= \mathcal{J}_F\{\frac{(\sqrt{g})_{,k}}{\sqrt{g}} - \Gamma^i_{ki}\}\mathbf{g}^k. \end{aligned} \tag{4.58}$$

This vanishes by (3.107).

References

Chadwick, P. (1976). *Continuum Mechanics: Consise Theory and Problems.* Dover, New York.

Doyle, T. C. and Ericksen, J. L. (1956). Nonlinear elasticity, in: *Advances in Applied Mechanics* IV. Academic Press, New York.

Flügge, W. (1972) *Tensor Analysis and Continuum Mechanics.* Springer, Berlin.

Green, A. E. and Adkins, J. E. (1970). *Large Elastic Deformations.* Oxford University Press, Oxford.

Green, A. E. and Zerna, W. (1968). *Theoretical Elasticity.* Oxford University Press, Oxford.

Sedov, L. I. (1966). *Foundations of the Non-linear Mechanics of Continua*, trans. by Schoenfeld-Reiner, R., ed. by Adkins, J. E., and Spencer, A. J. M. Pergamon Press, Oxford.

5

Elastic and plastic deformations

Recall the observation made in Chapter 1 to the effect that the elastic properties of a crystalline solid are largely insensitive to plastic deformation. This is a manifestation of the fact that plastic deformation is accommodated by the relative slip of adjacent crystallographic planes while the underlying crystalline lattice remains intact. This means that there exists an undistorted state of the material—the seat of the crystalline symmetry—that evolves with plastic deformation. We attempt to codify this notion in the present chapter, and thereby to construct the rudiments of a theory of elastic-plastic deformation.

5.1 Elastic–plastic deformation, dislocation density

With reference to the discussion in Section 2.4, let $g_{\kappa_i(p)}$ be the symmetry group relative to a local undistorted state $\kappa_i(p)$ of the material point p; thus, $g_{\kappa_i(p)} \subset Orth^+$. This is meaningful whether or not the solid is crystalline. We regard $\kappa_i(p)$ as an evolving local configuration in accordance with the foregoing comments. To acknowledge conventional terminology, the subscript on $\kappa_i(p)$ stands for "local *intermediate* configuration."

Noll's rule—see (2.100)—then yields the symmetry group at the material point p in a global reference configuration κ,

$$g_{\kappa(p)} = \{\mathbf{KRK}^{-1} \mid \mathbf{R} \in g_{\kappa_i(p)}\}, \tag{5.1}$$

where $\mathbf{K}(\mathbf{x}, t)$, in which $\mathbf{x} = \boldsymbol{\kappa}(p)$ the position of p in κ, is the map from the local configuration $\kappa_i(p)$ to the tangent space $T_\kappa(p)$ at p. Recalling the discussion surrounding Noll's rule in Section 2.4, $\mathbf{K}(\mathbf{x}, t)$ is in general *not* a gradient field. The inverse map,

$$\mathbf{G} = \mathbf{K}^{-1}, \tag{5.2}$$

taking $T_\kappa(p)$ to $\kappa_i(p)$, is commonly called the *plastic deformation*. It is the generalization to three dimensions of the plastic stretch λ_p discussed in Chapter 1, and is typically denoted by $\mathbf{F}_p$ in most of the literature.

A Course on Plasticity Theory. David J. Steigmann, Oxford University Press. © David J. Steigmann (2022).
DOI: 10.1093/oso/9780192883155.003.0005

Note that we are tacitly interpreting $\kappa_i(p)$ as a vector space rather than a configuration in the usual sense. Later, we will interpret $\kappa_i(p)$ as the tangent space to a certain non-Euclidean manifold associated with the material point p.

In the same way we can contemplate a map $\mathbf{H}$ from $\kappa_i(p)$ to the tangent space $T\kappa_t(p)$, at p, to the current global configuration κ_t of the body B. This map is the generalization of the elastic stretch λ_e mentioned in Chapter 1. It represents the elastic distortion of the lattice, *at* the point p, required to place it into κ_t, and is more often denoted by $\mathbf{F}_e$. Identifying p with $\mathbf{x} = \boldsymbol{\kappa}(p)$ as before, we denote this *elastic deformation* by $\mathbf{H}(\mathbf{x}, t)$. Being defined on $\kappa_i(p)$ rather than on a neighborhood of p in an actual configuration, this too is generally not a gradient field.

Of course there is another map, $\mathbf{F}(\mathbf{x}, t)$, from $T\kappa(p)$ to $T\kappa_t(p)$. This is the deformation *gradient* $\nabla\boldsymbol{\chi}(\mathbf{x}, t)$, the gradient at $\mathbf{x}$ of the deformation field $\boldsymbol{\chi}(\cdot, t)$ mapping the global configuration κ to the global configuration κ_t. Because $T\kappa(p)$ is the result of mapping $\kappa_i(p)$ by $\mathbf{K}$, whereas $T\kappa_t(p)$ is the map of $\kappa_i(p)$ by $\mathbf{H}$, it follows that

$$\mathbf{H} = \mathbf{FK}, \quad \text{where} \quad \mathbf{F} = \nabla\boldsymbol{\chi}. \tag{5.3}$$

This, of course, is equivalent to $\mathbf{F} = \mathbf{HG}$ —just another notation for the famous decomposition $\mathbf{F} = \mathbf{F}_e\mathbf{F}_p$ —generalizing the formula $\lambda = \lambda_e\lambda_p$ of Chapter 1. Note, however, that whereas it is true that $\lambda = \lambda_p\lambda_e$, it is *not* true that $\mathbf{F} = \mathbf{GH}$, for the simple reason that tensor multiplication does not commute. An alternative explanation is that $T\kappa_t(p)$ is the range of $\mathbf{H}$, but not the domain of $\mathbf{G}$, implying that the proposed product is not well defined. As usual we suppose that κ is occupiable, i.e., that $\mathcal{J}_F > 0$. Assuming $\kappa_i(p)$ to be occupiable, too, we also have $\mathcal{J}_K > 0$ and $\mathcal{J}_H > 0$, the former implying that $\mathcal{J}_G(= \mathcal{J}_K^{-1}) > 0$.

To better understand what's going on in this decomposition we note that the undistorted state to which we have repeatedly referred is the state used by crystallographers to characterize crystal symmetry. This is normally regarded as a state of minimum energy of the atomic arrangement constituting the crystal in which the lattice is unstressed. However, a local undistorted state $\kappa_i(p)$ defined by the condition $g_{\kappa_i(p)} \subset Orth^+$ may support a state of stress whose form is dictated by its symmetry. For example, cubic crystals and isotropic materials may support a pure pressure while occupying an undistorted state. It is common, however, to assume that $\kappa_i(p)$ is stress free, to promote the analogy to the stress-free configuration of the uniaxial bar of Figure 1.1, achieved by unloading from a state that has undergone prior plastic deformation. However, it is not generally possible to have a state of vanishing stress at all points of the body. Usually there is a distribution of *residual stress* due to the presence of various defects such as the dislocations accompanying plastic slip. These induce local lattice distortions in the case of crystalline metals, for example, which in turn generate elastic strain and a consequent distribution of stress (Figure 5.1). This is typically the case even when the body is entirely unloaded, i.e., when no body forces are applied and the boundary tractions vanish.

Nevertheless it is possible, in principle, to remove the *mean* stress via an *equilibrium* unloading process. In particular, in equilibrium the mean value, $\bar{\mathbf{T}}$, of the Cauchy stress

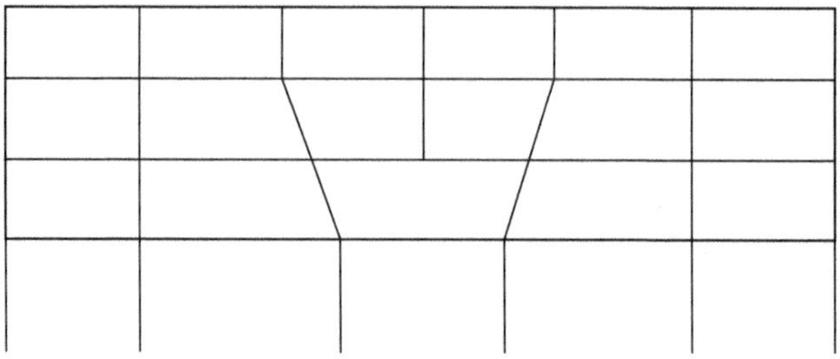

Figure 5.1 *Local lattice distortion in the vicinity of an edge dislocation in a cubic crystal.*

$\bar{\mathbf{T}}$ is given by

$$vol(\kappa_t)\bar{\mathbf{T}} = Sym\{\int_{\partial\kappa_t} \mathbf{t} \otimes \mathbf{y} da + \int_{\kappa_t} \rho \mathbf{b} \otimes \mathbf{y} dv\}, \tag{5.4}$$

where $\mathbf{t}$ and $\mathbf{b}$, respectively, are the boundary traction and body force and $\mathbf{y}$ is the position of a material point in the current configuration κ_t of the body. See Chadwick's book for a simple derivation based on (1.32). Accordingly, $\bar{\mathbf{T}} = \mathbf{0}$ if the entire body is unloaded.

The mean-value theorem for continuous functions guarantees the existence of $\bar{\mathbf{y}} \in \kappa_t$ such that $\mathbf{T}(\bar{\mathbf{y}}, t) = \bar{\mathbf{T}}$. Therefore, $\mathbf{T}(\bar{\mathbf{y}}, t) = \mathbf{0}$ for some $\bar{\mathbf{y}} \in \kappa_t$ if the body is unloaded and in equilibrium. Let

$$d(\kappa_t) = \sup_{\mathbf{y},\mathbf{z}\in\kappa_t} |\mathbf{y} - \mathbf{z}|. \tag{5.5}$$

This is the *diameter* of κ_t. Then for every $\mathbf{y} \in \kappa_t$ we have $\mathbf{T}(\mathbf{y}, t) \to \mathbf{T}(\bar{\mathbf{y}}, t)$ as $d(\kappa_t) \to 0$. Accordingly, the local value of the stress is made arbitrarily close to zero as the diameter of the body shrinks to zero.

Of course, it is not possible to reduce the diameter of a given body to zero. However, we may regard any body as the union of an arbitrary number of arbitrarily small disjoint sub-bodies $\pi_t^{(n)}$, i.e., $\kappa_t = \cup_{n=1}^{\infty}\pi_t^{(n)}$, with $d(\pi_t^{(n)}) \to 0$. Imagine separating these sub-bodies and unloading them individually. We then have $\mathbf{T}(\mathbf{y}, t) \to \mathbf{0}$ for every $\mathbf{y} \in \pi_t^{(n)}$, for every n. Because every $\mathbf{y}$ in κ_t belongs to some $\pi_t^{(n)}$, this process results in a state in which the material is destressed pointwise. We expect that each piece $\pi_t^{(n)}$ will experience some distortion in this process, and therefore that the unstressed sub-bodies cannot be made congruent to fit together into a connected region of three-dimensional Euclidean space. Thus, there is no *global* stress-free configuration of the body, and hence no position field $\mathbf{r}$, say, such that $d\mathbf{r} = \mathbf{G}d\mathbf{x}$ (or $\mathbf{H}^{-1}d\mathbf{y}$); that is, there is no neighborhood in the vanishingly small unloaded sub-bodies that can be used to define a gradient of a position field. Accordingly, as we have already noted, neither **G** nor **H** is a gradient. It follows that for any closed curve $\Gamma \subset \kappa$ the vector

$$\mathbf{B}_\Gamma = \int_\Gamma \mathbf{G}d\mathbf{x} \tag{5.6}$$

does not vanish. This is called the *Burgers vector* associated with Γ. In the same way we can define

$$\mathbf{b}_\gamma = \int_\gamma \mathbf{H}^{-1}d\mathbf{y}, \tag{5.7}$$

not to be confused with the body force, where $\gamma = \chi(\Gamma, t)$ is the image in κ_t of Γ under the deformation map.

Problem 5.1 Show that if the tractions acting on the disjoint sub-bodies $\pi_t^{(n)}$ vanish, then they become stress free in the limit $d(\pi_t^{(n)}) \to 0$ whether or not they are in equilibrium, provided that the norm of $\rho(\mathbf{b} - \dot{\mathbf{v}})$, where $\mathbf{b}$ is the body force, is bounded.

Problem 5.2 Show that the Burgers vectors $\mathbf{b}_\gamma$ and $\mathbf{B}_\Gamma$ are one and the same.

Consider a simply connected surface $\Omega \subset \kappa$. On combining (5.6) with Stokes' formula (3.120), applied in κ, we have the associated Burgers vector

$$\mathbf{B}(\Omega, t) = \int_\Omega \boldsymbol{\alpha}_\mathcal{K}^t \boldsymbol{\nu} dA, \tag{5.8}$$

where $\boldsymbol{\nu}$ is a unit normal to Ω and

$$\boldsymbol{\alpha}_\mathcal{K} = Curl\mathbf{K}^{-1} \tag{5.9}$$

is the *referential dislocation density* tensor in which *Curl* is the curl based on $\mathbf{x} \in \kappa$. This presumes the plastic deformation to be smooth and hence continuous. We will revisit this formula for discontinuous plastic deformation fields in Chapter 6.

In the same way we have

$$\mathbf{B}(\Omega, t) = \mathbf{b}(\omega, t) = \int_\omega \boldsymbol{\alpha}_{\mathcal{K}_t}^t \mathbf{n}da, \tag{5.10}$$

where $\omega = \chi(\Omega, t)$ is the image of Ω in κ_t with unit-normal field $\mathbf{n}$, related to $\boldsymbol{\nu}$ by the Piola–Nanson formula (1.35), and

$$\boldsymbol{\alpha}_{\mathcal{K}_t} = curl\mathbf{H}^{-1} \tag{5.11}$$

is the *spatial dislocation density* tensor in which *curl* is based on $\mathbf{y} \in \kappa_t$.

The first equality in (5.10) suggests that these dislocation densities are not independent. We establish their relationship via a somewhat circuitous argument. Thus consider two global reference configurations κ_1 and κ_2, say. Let $\boldsymbol{\lambda}$ be an invertible differentiable map from the first to the second, so that the position of a material point p in κ_2 is $\mathbf{x}_2 = \boldsymbol{\lambda}(\mathbf{x}_1)$, where $\mathbf{x}_1$ is its position in κ_1. Let $\mathbf{R} = \nabla_1\boldsymbol{\lambda}$ be its gradient with respect to $\mathbf{x}_1$. We hold κ_t and $\kappa_i(p)$ fixed in this discussion, so that $\mathbf{H}$ is unaffected. Then, because $\mathbf{K}^{-1}d\mathbf{x} = \mathbf{H}^{-1}d\mathbf{y}$ in which the right-hand side is fixed, it follows, using obvious notation, that

$$\mathbf{K}_1^{-1}d\mathbf{x}_1 = \mathbf{K}_2^{-1}d\mathbf{x}_2. \tag{5.12}$$

From $d\mathbf{x}_2 = \mathbf{R}d\mathbf{x}_1$ we then have

$$\mathbf{K}_2 = \mathbf{R}\mathbf{K}_1. \tag{5.13}$$

If Ω_1 is a simply connected surface in κ_1, then, with $\Omega_2 = \boldsymbol{\lambda}(\Omega_1)$,

$$\int_{\Omega_2}(Curl_2\mathbf{K}_2^{-1})^t\boldsymbol{\nu}_2 dA_2 = \int_{\partial\Omega_2}\mathbf{K}_2^{-1}d\mathbf{x}_2 = \int_{\partial\Omega_1}\mathbf{K}_1^{-1}d\mathbf{x}_1 = \int_{\Omega_1}(Curl_1\mathbf{K}_1^{-1})^t\boldsymbol{\nu}_1 dA_1. \tag{5.14}$$

The outer equality may be reduced, with the aid of the Piola–Nanson formula

$$\boldsymbol{\nu}_2 dA_2 = \mathbf{R}^*\boldsymbol{\nu}_1 dA_1, \tag{5.15}$$

to

$$\int_{\Omega_1}(Curl_2\mathbf{K}_2^{-1})^t\mathbf{R}^*\boldsymbol{\nu}_1 dA_1 = \int_{\Omega_1}(Curl_1\mathbf{K}_1^{-1})^t\boldsymbol{\nu}_1 dA_1. \tag{5.16}$$

Because Ω_1 is arbitrary—and hence also $\boldsymbol{\nu}_1$ —it follows that

$$(Curl_2\mathbf{K}_2^{-1})^t\mathbf{R}^* = (Curl_1\mathbf{K}_1^{-1})^t, \tag{5.17}$$

which may be rearranged as

$$J_R Curl_2\mathbf{K}_2^{-1} = \mathbf{R}(Curl_1\mathbf{K}_1^{-1}); \tag{5.18}$$

equivalently,

$$J_{K_2}\mathbf{K}_2^{-1}Curl_2\mathbf{K}_2^{-1} = J_{K_1}\mathbf{K}_1^{-1}Curl_1\mathbf{K}_1^{-1}. \tag{5.19}$$

It follows that

$$\boldsymbol{\alpha} = J_K\mathbf{K}^{-1}Curl\mathbf{K}^{-1} \tag{5.20}$$

furnishes a measure of dislocation density that is intrinsic in the sense of being insensitive to the reference configuration. For this reason it is often called the *true dislocation density.*

Alternatively, we may proceed from (5.10), or use a formula like (5.18), with obvious notational adjustments, to establish that

$$\boldsymbol{\alpha} = J_H \mathbf{H}^{-1} curl \mathbf{H}^{-1}. \tag{5.21}$$

Following essentially the same reasoning as that leading to (5.19), we conclude that $\boldsymbol{\alpha}$ is insensitive to arbitrary invertible differentiable variations of κ_t, and hence that it is invariant under superposed rigid-body deformations in particular. Finally, using (5.9), (5.11), (5.20), and (5.21) yields the relation

$$J_F \boldsymbol{\alpha}_{\mathcal{K}_t} = \mathbf{F}\boldsymbol{\alpha}_{\mathcal{K}} \tag{5.22}$$

connecting the spatial and referential dislocation densities. It follows from this that $\boldsymbol{\alpha}_{\mathcal{K}_t}$ is actually a two-point tensor.

We have noted that $\kappa_i(p)$ should be interpreted as a vector space. It may be regarded as the tangent space to a certain body manifold, but this manifold is not Euclidean as it does not support a position field. This interpretation is the basis of an elegant differential-geometric theory of plastically deformed bodies, which we pause to outline in some detail.

5.2 Differential-geometric considerations

As in Chapter 4 we use convected coördinates ξ^i to identify the material point p. Recall that these induce the linearly independent and positively oriented coordinate tangent basis vectors $\mathbf{e}_i = \mathbf{x}_{,i}$ in $T_{\mathcal{K}}$. We use these with the plastic deformation tensor $\mathbf{G}$ to define vectors $\mathbf{m}_k \in \kappa_i(p)$ by

$$\mathbf{m}_i = \mathbf{G}\mathbf{e}_i. \tag{5.23}$$

These too are linearly independent and positively oriented because J_G is positive; thus, $\{\mathbf{m}_i\}$ is a basis for $\kappa_i(p)$. Then, as in (4.9),

$$\mathbf{G} = \mathbf{m}_i \otimes \mathbf{e}^i, \tag{5.24}$$

and the *plastic* Cauchy–Green deformation tensor is

$$\mathbf{G}^t\mathbf{G} = m_{ij}\mathbf{e}^i \otimes \mathbf{e}^j, \quad \text{where} \quad m_{ij} = \mathbf{m}_i \cdot \mathbf{m}_j \tag{5.25}$$

is the metric induced in $\kappa_i(p)$ by the coordinates. As this is positive definite, so too is the matrix (m_{ij}). Accordingly $m > 0$, where $m = \det(m_{ij})$, and we have well-defined reciprocal basis elements $\mathbf{m}^i = m^{ij}\mathbf{m}_j$, where $(m^{ij}) = (m_{ij})^{-1}$.

With these in hand we define the Weitzenböck connection

$$\hat{\Gamma}_{ij}^{k} = \mathbf{m}^k \cdot \mathbf{m}_{i,j}, \tag{5.26}$$

as in (3.138). Proceeding exactly as in the passage from the latter to (3.142), we conclude that

$$\hat{R}^k_{\cdot mlj} = 0, \tag{5.27}$$

where

$$\hat{R}^k_{\cdot mlj} = 2(\hat{\Gamma}^k_{m[j,l]} - \hat{\Gamma}^k_{i[j}\hat{\Gamma}^k_{|m|l]}) \tag{5.28}$$

is the associated curvature. Thus, the geometry with connection $\hat{\Gamma}$ is flat. The vanishing of the curvature tensor constitutes the integrability condition for the system

$$\mathbf{m}_{i,j} = \hat{\Gamma}_{ij}^{k}\mathbf{m}_k, \tag{5.29}$$

which follows from (5.26), ensuring the existence of the fields $\mathbf{m}_i$ in any path-connected domain, this of course being an immediate consequence of (5.23). Further, $(5.25)_2$ and (5.29) imply that $\hat{\Gamma}_{ij}^{k}$ is compatible with the metric m_{ij} in the sense that

$$m_{ij,k} = \hat{\Gamma}_{jik} + \hat{\Gamma}_{ijk}, \quad \text{where} \quad \hat{\Gamma}_{jik} = m_{lj}\hat{\Gamma}^{l}_{ik}. \tag{5.30}$$

This is Ricci's lemma. That is, the covariant derivative of the metric with respect to the connection vanishes—see (3.94).

The torsion induced by this connection is

$$\hat{T}^k_{\cdot ij} = \hat{\Gamma}^k_{[ij]} = \mathbf{m}^k \cdot \mathbf{m}_{[i,j]}, \tag{5.31}$$

where

$$\begin{aligned} \mathbf{m}_{i,j} &= \mathbf{G}_{,j}\mathbf{e}_i + \mathbf{G}\mathbf{e}_{i,j} = \mathbf{G}_{,j}\mathbf{e}_i + \bar{\Gamma}^k_{ij}\mathbf{G}\mathbf{e}_k \\ &= \mathbf{G}_{,j}\mathbf{e}_i + \bar{\Gamma}^k_{ij}\mathbf{m}_k, \end{aligned} \tag{5.32}$$

in which $\bar{\Gamma}$ is the (symmetric) Levi-Civita connection induced by the coordinates in κ. Thus, $\mathbf{m}_{[i,j]} = \mathbf{G}_{,[j}\mathbf{e}_{i]}$, and as this is generally non-zero, it follows that the torsion does not vanish; the geometry based on $\hat{\Gamma}$ is not Euclidean.

Exceptionally, if $\mathbf{G}$ were a smooth gradient field, then we would have $\mathbf{m}_i = \mathbf{r}_{,i}$ for some twice differentiable field $\mathbf{r}$, say. In this event we would have

$$\mathbf{G}_{,j} = (\mathbf{r}_{,kj} - \bar{\Gamma}^l_{kj}\mathbf{r}_{,l}) \otimes \mathbf{e}^k, \tag{5.33}$$

giving $\mathbf{m}_{[i,j]} = \mathbf{r}_{,[ij]}$, which vanishes identically, together with the torsion. The underlying geometry would then be Euclidean. This subsumes homogeneous plastic deformations for which $\mathbf{G}$ is spatially uniform, in which case (5.29) and (5.32) imply that $\hat{\Gamma} = \bar{\Gamma}$.

In view of (5.30) we can apply (3.99) to the metric m_{ij} and connection $\hat{\Gamma}^k_{ij}$ to obtain

$$\tfrac{1}{2}(m_{ki,j} + m_{kj,i} - m_{ij,k}) = m_{li}\hat{\Gamma}^l_{[kj]} + m_{lj}\hat{\Gamma}^l_{[ki]} + m_{lk}\hat{\Gamma}^l_{(ij)}. \tag{5.34}$$

Let $\{^{\,m}_{ij}\}(= \{^{\,m}_{ji}\})$ be the Levi-Civita connection based on the same metric, i.e.,

$$\{^{\,m}_{ij}\} = \tfrac{1}{2}m^{mk}(m_{ki,j} + m_{kj,i} - m_{ij,k}). \tag{5.35}$$

The curly bracket is a rather old-fashioned notation for the connection. We resurrect it here to avoid introducing too many Γ's. Equation (5.34) is then equivalent to

$$\{^{\,m}_{ij}\} = (\hat{T}_{\cdot kj}^{\;\,l} m_{li} + \hat{T}_{\cdot ki}^{\;\,l} m_{lj})m^{mk} + \hat{\Gamma}^m_{(ij)}. \tag{5.36}$$

Solving for $\hat{\Gamma}^m_{(ij)}$ and substituting into the identity $\hat{\Gamma}^m_{ij} = \hat{\Gamma}^m_{(ij)} + \hat{\Gamma}^m_{[ij]} = \hat{\Gamma}^m_{(ij)} + \hat{T}_{\cdot ij}^{\;\,m}$, we conclude that

$$\hat{\Gamma}^k_{ij} = \{^{\,k}_{ij}\} - K_{\cdot ij}^{\;\,k}, \tag{5.37}$$

where

$$K_{\cdot ij}^{\;\,k} = (\hat{T}_{\cdot nj}^{\;\,l} m_{li} + \hat{T}_{\cdot ni}^{\;\,l} m_{lj})m^{kn} - \hat{T}_{\cdot ij}^{\;\,k} \tag{5.38}$$

is the *contortion* tensor. Clearly $K_{\cdot ij}^{\;\,k}$ vanishes if $\hat{T}_{\cdot ij}^{\;\,k}$ vanishes. Conversely, if $K_{\cdot ij}^{\;\,k}$ vanishes, then $\hat{T}_{\cdot ji}^{\;\,k}\ (= K_{\cdot [ij]}^{\;\,k})$ vanishes.

The torsion $\hat{T}_{\cdot ij}^{\;\,k}$ is equivalent to the true dislocation density, which may be written in the form

$$\boldsymbol{\alpha} = J_G^{-1}\mathbf{G}\,Curl\mathbf{G}. \tag{5.39}$$

To establish this recall that the curl is defined by

$$(Curl\mathbf{G})\mathbf{c} = Curl(\mathbf{G}^t\mathbf{c}), \tag{5.40}$$

for any *fixed* vector **c**. Writing $\mathbf{c} = c_i\mathbf{m}^i$ and using (5.24) in the form $\mathbf{G}^t = \mathbf{e}^i \otimes \mathbf{m}_i$, we have $\mathbf{G}^t\mathbf{c} = c_i\mathbf{e}^i$, and the referential form of (3.117) yields

$$Curl(\mathbf{G}^t\mathbf{c}) = \bar{\varepsilon}^{ijk} c_{j|i}\mathbf{e}_k, \tag{5.41}$$

where $c_{j|i} = c_{j,i} - c_l\bar{\Gamma}^l_{ji}$ and $\bar{\varepsilon}^{ijk}(= e^{ijk}/\sqrt{e})$ are the components of the referential permutation tensor. Because $\bar{\varepsilon}^{ijk} = \bar{\varepsilon}^{[ij]k}$, whereas $\bar{\Gamma}^l_{ji} = \bar{\Gamma}^l_{ij}$, this simplifies to

$$Curl(\mathbf{G}^t\mathbf{c}) = \bar{\varepsilon}^{ijk} c_{[j,i]}\mathbf{e}_k = \bar{\varepsilon}^{ijk}\mathbf{e}_k(\mathbf{m}_{[j,i]} \cdot \mathbf{c}) = (\bar{\varepsilon}^{ijk}\mathbf{e}_k \otimes \mathbf{m}_{[j,i]})\mathbf{c}, \tag{5.42}$$

and (5.40) furnishes

$$Curl\mathbf{G} = \bar{\varepsilon}^{ijk}\mathbf{e}_k \otimes \mathbf{m}_{[j,i]}. \tag{5.43}$$

Then (5.29) and (5.39) combine to give

$$\boldsymbol{\alpha} = \alpha^{ij}\mathbf{m}_i \otimes \mathbf{m}_j, \tag{5.44}$$

where

$$\alpha^{ij} = \hat{\varepsilon}^{kli}\hat{T}^{\,j}_{\cdot lk}, \tag{5.45}$$

in which $\hat{\varepsilon}^{ijk} = e^{ijk}/\sqrt{m}$, with $m = \det(m_{ij})$, and use has been made of $\mathcal{J}_G = \sqrt{m/e}$.

Problem 5.3. Use the relevant version of (3.124) to invert (5.45), and thus show that

$$\hat{T}^{\,l}_{\cdot nm} = \tfrac{1}{2}\hat{\varepsilon}_{mnk}\alpha^{kl}.$$

These results make clear the fact that $\boldsymbol{\alpha}$ is equivalent to the torsion and reinforce our earlier conclusion that it is intrinsic; that is, independent of κ_t or κ.

5.3 Incompatibility of the elastic strain

The *elastic* strain is

$$\hat{\mathbf{E}} = \tfrac{1}{2}(\mathbf{H}^t\mathbf{H} - \mathbf{I}), \tag{5.46}$$

where, from (4.9), (5.2), and (5.24),

$$\mathbf{H} = \mathbf{F}\mathbf{K} = (\mathbf{g}_i \otimes \mathbf{e}^i)(\mathbf{e}_j \otimes \mathbf{m}^j) = \delta^i_j\mathbf{g}_i \otimes \mathbf{m}^j = \mathbf{g}_i \otimes \mathbf{m}^i. \tag{5.47}$$

Thus,

$$\hat{\mathbf{E}} = \hat{E}_{ij}\mathbf{m}^i \otimes \mathbf{m}^j, \tag{5.48}$$

where

$$\hat{E}_{ij} = \tfrac{1}{2}(g_{ij} - m_{ij}). \tag{5.49}$$

We intend to show that this strain is incompatible if, as is typically the case, the torsion $\hat{T}^k_{\cdot ij}$ is non-zero. This means that elastic strain cannot vanish in the presence of a density of dislocations. Because elastic strain gives rise to stress, as we shall see in Chapter 6, it follows that dislocations indirectly generate a stress field. Thus, a *residual stress* field persists even in the absence of any loading applied to the body. To demonstrate this we solve (5.49) for the metric m_{ij}, substitute into (5.37), and impose the flatness conditions $\hat{R}^k_{\cdot mlj} = 0$ and $R^k_{\cdot mlj} = 0$. Here R is the Riemann curvature of Euclidean space based on the Levi-Civita connection Γ, computed from the metric g_{ij} and induced in κ_t by the coordinates. However, before proceeding we dispose of some preliminaries.

First, we recall the condition (3.137) satisfied by all curvature tensors. In particular,

$$\hat{R}_{pm(lj)} = 0, \tag{5.50}$$

where—see (3.97) and (4.30)—

$$\hat{R}_{pmlj} = m_{pk}\hat{R}^k_{\cdot mlj} = \hat{\Gamma}_{pmj,l} - \hat{\Gamma}_{pml,j} + \hat{\Gamma}^i_{ml}\hat{\Gamma}_{ipj} - \hat{\Gamma}^i_{mj}\hat{\Gamma}_{ipl}. \tag{5.51}$$

In fact, we also have

$$\hat{R}_{(pm)lj} = 0. \tag{5.52}$$

Proof: It follows from (5.51) that

$$\begin{aligned} \hat{R}_{pmlj} + \hat{R}_{mplj} &= \hat{\Gamma}_{pmj,l} + \hat{\Gamma}_{mpj,l} - (\hat{\Gamma}_{pml,j} + \hat{\Gamma}_{mpl,j}) \\ &\quad + \hat{\Gamma}^i_{ml}\hat{\Gamma}_{ipj} + \hat{\Gamma}^i_{pl}\hat{\Gamma}_{imj} - (\hat{\Gamma}^i_{mj}\hat{\Gamma}_{ipl} + \hat{\Gamma}^i_{pj}\hat{\Gamma}_{iml}). \end{aligned} \tag{5.53}$$

But $\hat{\Gamma}^i_{pl}\hat{\Gamma}_{imj} = m_{in}\hat{\Gamma}^n_{mj}\hat{\Gamma}^i_{pl} = \hat{\Gamma}_{npl}\hat{\Gamma}^n_{mj} = \hat{\Gamma}^i_{mj}\hat{\Gamma}_{ipl}$, implying that the second line vanishes. Using (5.30) we then reduce this to

$$\hat{R}_{(pm)lj} = m_{mp,[jl]}, \tag{5.54}$$

and thus establish (5.52), assuming the metric to be a twice-differentiable function.

Taken together with (5.50), this implies that

$$\hat{R}_{pmlj} = \hat{R}_{[pm][lj]}. \tag{5.55}$$

We can then proceed as in the derivation of (4.33) and (4.34) to conclude that $\hat{R}$ stands in one-to-one relation to $\hat{\Pi}$, defined by

$$\hat{\Pi}^{ij} = \tfrac{1}{4}\hat{\epsilon}^{ikl}\hat{\epsilon}^{jmn}\hat{R}_{klmn}, \tag{5.56}$$

with inverse

$$\hat{R}_{pqrs} = \hat{\epsilon}_{ipq}\hat{\epsilon}_{jrs}\hat{\Pi}^{ij}, \tag{5.57}$$

where $\hat{\epsilon}_{ijk} = \sqrt{m}e_{ijk}$. Accordingly, (5.27) is equivalent to

$$\hat{\Pi}^{ij} = 0. \tag{5.58}$$

However, recalling that the derivation of the major symmetry condition (4.32) relied on a Levi-Civita connection, it would appear that we do not have $\hat{R}_{pmlj} = \hat{R}_{ljpm}$ because $\hat{\Gamma}^{i}_{jk}$ is not Levi-Civita. Apparently, then, $\hat{\Pi}^{ij} \neq \hat{\Pi}^{ji}$. But, in fact, it turns out that $\hat{\Pi}^{[ij]} = 0$, implying that (5.58) represents only six restrictions, as in the discussion (in Section 4.2) about compatibility of the overall strain.

Problem 5.4 Prove the identity $Div[(Curl\mathbf{A})^t] = \mathbf{0}$, and establish the equivalence of the following statements:

(a) $Div[(Curl\mathbf{G})^t] = \mathbf{0}$, where $\mathbf{G}$ is the plastic part of the deformation gradient.

(b) $\hat{\epsilon}^{ijk}(\hat{T}^{n}_{.ji,k} + \hat{T}^{l}_{.ji}\hat{\Gamma}^{n}_{lk}) = 0$, where $\hat{\epsilon}$, $\hat{T}$, and $\hat{\Gamma}$, respectively, are the permutation tensor, the torsion tensor, and the connection in the intermediate state.

(c) $\hat{\epsilon}^{mlj}\hat{R}_{kmlj} = 0$, where $\hat{R}$ is the curvature tensor based on $\hat{\Gamma}$.

(d) $\hat{\Pi}^{[ij]} = 0$.

Thus, the equations in (b)–(d) are identities.

Returning to the matter at hand, we follow a procedure similar to that used in Section 4.2 and define a tensor with components

$$\hat{S}^{k}_{\cdot ij} = \Gamma^{k}_{ij} - \{^{\,k}_{ij}\}. \tag{5.59}$$

Then (5.37) may be expressed in the form

$$\hat{\Gamma}^{k}_{ij} = \Gamma^{k}_{ij} - L^{k}_{\cdot ij}, \quad \text{where} \quad L^{k}_{\cdot ij} = \hat{S}^{k}_{\cdot ij} + K^{k}_{\cdot ij}, \tag{5.60}$$

and (3.136) yields

$$\begin{aligned}\hat{R}^{k}_{\cdot mlj} &= R^{k}_{\cdot mlj} + \Gamma^{k}_{ij}L^{i}_{\cdot ml} - \Gamma^{k}_{il}L^{i}_{\cdot mj} + L^{k}_{\cdot ij}\Gamma^{i}_{ml} - L^{k}_{\cdot il}\Gamma^{i}_{mj} \\ &\quad + L^{k}_{\cdot ml,j} - L^{k}_{\cdot mj,l} + L^{k}_{\cdot il}L^{i}_{\cdot mj} - L^{k}_{\cdot ij}L^{i}_{\cdot ml},\end{aligned} \tag{5.61}$$

where R is the Riemann tensor in κ_t. Recalling the geometer's trick described in Section 4.2, we choose the convected coordinates to be Cartesian in κ_t. This is always possible at any *fixed* instant t, though, of course, the coordinates, insofar as they convect with the deformation, will generally not be Cartesian at earlier and later instants. This choice means that the Γ's vanish instantaneously. Because $\hat{R}$ and R vanish at all times, and hence at the instant in question, (5.61) reduces to

$$L^{k}_{\cdot ml,j} - L^{k}_{\cdot mj,l} + L^{k}_{\cdot il}L^{i}_{\cdot mj} - L^{k}_{\cdot ij}L^{i}_{\cdot ml} = 0. \tag{5.62}$$

Reverting to general coordinates we then have the tensor equation

$$L^{k}_{\cdot ml;j} - L^{k}_{\cdot mj;l} + L^{k}_{\cdot il}L^{i}_{\cdot mj} - L^{k}_{\cdot ij}L^{i}_{\cdot ml} = 0, \tag{5.63}$$

where the semi-colon is the covariant derivative based on the connection Γ. This is valid at all times because the coordinates are no longer required to be Cartesian.

Problem 5.5 Establish the formula

$$\hat{S}^{k}_{\cdot ij} = m^{km}(\hat{E}_{mi;j} + \hat{E}_{mj;i} - \hat{E}_{ij;m}).$$

This result may be combined with $(5.60)_2$ and (5.63) to arrive at a set of differential constraints involving the elastic strain $\hat{E}$, the metric m, and the contortion K, which generally preclude the possibility that the elastic strain vanishes identically in the body. To see this we investigate the implications of the assumption that the $\hat{E}_{ij}$ vanish together with their derivatives. Then the $\hat{S}^{k}_{\cdot ij}$ vanish and the $L^{k}_{\cdot ij}$ coincide with $K^{k}_{\cdot ij}$, reducing (5.63) to

$$K^{k}_{m[j,l]} - K^{k}_{i[j}K^{i}_{|m|l]} = 0, \tag{5.64}$$

where the coordinates have again been taken to be Cartesian in κ_t , with $\Gamma^{k}_{ij} = 0$, for the sake of simplicity. This, in turn, yields the three restrictions

$$\hat{\varepsilon}^{jml}(K^{k}_{mj,l} - K^{k}_{ij}K^{i}_{ml}) = 0. \tag{5.65}$$

Compare this to the identity proved in Problem 5.4(b), which implies, for vanishing $\hat{S}^{k}_{\cdot ij}$ and again with $\Gamma^{k}_{ij} = 0$, that

$$\hat{\varepsilon}^{jml}(\hat{T}^{k}_{mj,l} - K^{k}_{ij}\hat{T}^{i}_{ml}) = 0. \tag{5.66}$$

If we add the last two equations, and note, from (5.38), that $K^i_{ml} + \hat{T}^i_{ml}$ is symmetric in the subscripts, we find that the resulting equation is identically satisfied. Thus, (5.65) is also an identity. However, there are further restrictions contained in (5.64)—equivalent to $\hat{\Pi}^{(ij)} = 0$—that are not accounted for by (5.65), and these impose additional constraints on the manner in which the torsion tensor can depend on the coordinates. In the general case there is no reason why these extra constraints should be satisfied. Thus, the premise is false: the elastic strain field cannot be identically zero in the presence of a general non-zero torsion field.

References

Cermelli, P., and Gurtin, M. E. (2001). On the characterization of geometrically necessary dislocations in finite plasticity. *J. Mech. Phys. Solids* 49, 1539–68.

Chadwick, P. (1976). *Continuum Mechanics: Concise Theory and Problems.* Dover, New York.

Gairola, B. K. D. (1979). Non-linear elastic problems. In: Nabarro, F. R. N. (Ed.), *Dislocations in Solids*, Vol. 1. North-Holland, Amsterdam.

Rajagopal, K. R., and Srinivasa, A. R. (1998). Mechanics of the inelastic behavior of materials—part 1: Theoretical underpinnings. *Int. J. Plasticity* 14, 945–67.

6

Energy, stress, dissipation, and plastic evolution

In this chapter we outline the basic constitutive theory for energy, stress, and the dissipation accompanying plastic evolution. We extend the notion of dislocation density to accommodate surface dislocation and discuss its role in relaxing the stress in adjoining crystal grains and at phase interfaces, and further results concerning the dissipation associated with the evolution of the latter are given. Also discussed is the notion of yield and the role it plays in the flow rule for plastic evolution, and the essential distinction between crystallinity and isotropy in the basic formulation of the theory.

6.1 Materially uniform elastic bodies

Our basic conceptual framework is the theory of materially uniform simple materials pioneered by Noll, which posits the existence of a strain-energy function, and associated stress, given by constitutive functions that are the same for all material points. This furnishes an appropriate model for a uniform single crystal, for example. Crystalline aggregates may then be viewed as composites consisting of individual materially uniform crystal grains. This is adequate for most applications involving metal plasticity, and also furnishes a foundation for theories of non-crystalline solids.

For example, with reference to (2.36) the strain energy contained in a part $\pi_t \subset \kappa_t$ is

$$\mathcal{U} = \int_{\pi_t} \psi(\mathbf{H})dv, \tag{6.1}$$

in which the constitutive function $\psi(\mathbf{H})$ for the energy per unit current volume is assumed to be the same function of $\mathbf{H}$ at all p, implying that any dependence on the latter occurs implicitly, via the field $\mathbf{H}(\mathbf{x}, t)$. While the restriction to uniform functions is not essential, what *is* essential is the assumption that energy arises in response to distortion of the crystal lattice, as represented by $\mathbf{H}$. This assumption effectively embodies G. I. Taylor's observations to the effect that elastic properties, encoded in the undistorted lattice, are insensitive to plastic deformation. We extend the same constitutive assumption to non-crystalline materials.

A Course on Plasticity Theory. David J. Steigmann, Oxford University Press. © David J. Steigmann (2022).
DOI: 10.1093/oso/9780192883155.003.0006

Proceeding as in (2.37), the strain energy per unit volume of a reference configuration κ is

$$\Psi = \mathcal{J}_F\psi. \tag{6.2}$$

It proves advantageous to introduce an energy W defined by

$$\Psi = \mathcal{J}_G W. \tag{6.3}$$

This may be viewed as the strain energy "per unit volume of $\kappa_i(p)$", although $\kappa_i(p)$ is a vector space rather than a configuration. That is, even though we cannot associate a volume with $\kappa_i(p)$, $\mathcal{J}_G$ is nevertheless well defined. Eliminating Ψ and using $\mathcal{J}_G^{-1}\mathcal{J}_F = \mathcal{J}_H$, we conclude that W is the function of $\mathbf{H}$ given by

$$W(\mathbf{H}) = \mathcal{J}_H\psi(\mathbf{H}). \tag{6.4}$$

Note that W plays the same role for $\mathbf{H}$ as that played by Ψ for $\mathbf{F}$ in Chapter 2, and in fact the two functions coincide when $\mathbf{K} = \mathbf{I}$, i.e., when $\mathbf{F} = \mathbf{H}$. Accordingly it will serve as our basic constitutive function. From (5.2) and (5.3), the energy per unit volume of κ is

$$\Psi(\mathbf{F},\mathbf{K}) = \mathcal{J}_K^{-1} W(\mathbf{FK}). \tag{6.5}$$

Evidently, then, at fixed t a plastic deformation field $\mathbf{K}(\mathbf{x})$ gives rise to a strain-energy function for a conventional *non-uniform* elastic solid of the kind encountered in Chapter 2. Recalling the requirement (2.24) that such functions be invariant under the replacement $\mathbf{F} \to \bar{\mathbf{Q}}\mathbf{F}$ for any rotation $\bar{\mathbf{Q}}$, we thus have

$$\mathcal{J}_K^{-1} W(\mathbf{FK}) = \Psi(\mathbf{F},\mathbf{K}) = \Psi(\bar{\mathbf{Q}}\mathbf{F},\mathbf{K}) = \mathcal{J}_K^{-1} W(\bar{\mathbf{Q}}\mathbf{FK}), \tag{6.6}$$

or simply

$$W(\mathbf{H}) = W(\bar{\mathbf{Q}}\mathbf{H}), \tag{6.7}$$

and this holds at all t as W does not depend explicitly on t. Then, exactly as in the derivation of (2.27),

$$W(\mathbf{H}) = U(\hat{\mathbf{E}}), \quad \text{where} \quad \hat{\mathbf{E}} = \tfrac{1}{2}(\mathbf{H}^t\mathbf{H} - \mathbf{I}) \tag{6.8}$$

is, of course, the elastic strain, and U is again an energy "per unit volume of $\kappa_i(p)$." We have put an overbar on the rotation $\bar{\mathbf{Q}}$ to distinguish it from the spatially uniform rotation $\mathbf{Q}(t)$ of (2.25) associated with a superposed rigid-body motion. This is because the argument leading to (2.24) is purely local, implying that $\bar{\mathbf{Q}}$ may depend on both $\mathbf{x}$ and t. This distinction will be discussed further in Section 6.4.

With these results in hand we may proceed as in Chapter 2 to derive the Piola stress

$$W_{\mathbf{H}} = \mathbf{H}\hat{\mathbf{S}} \tag{6.9}$$

in terms of the Piola–Kirchhoff stress

$$\hat{\mathbf{S}} = U_{\hat{\mathbf{E}}}, \tag{6.10}$$

both referred to $\kappa_i(p)$. With reference to (1.38), the first of these is related to the Cauchy stress $\mathbf{T}$ by

$$W_{\mathbf{H}} = \mathbf{T}\mathbf{H}^* = \mathbf{T}(\mathbf{F}\mathbf{K})^* = \mathbf{T}\mathbf{F}^*\mathbf{K}^* = \mathbf{P}\mathbf{K}^*, \tag{6.11}$$

where $\mathbf{P}$ is the Piola stress referred to κ, given, as in (2.10), by the partial derivative

$$\mathbf{P} = \Psi_{\mathbf{F}}, \tag{6.12}$$

at fixed $\mathbf{K}$. Thus,

$$W_{\mathbf{H}} = (\Psi_{\mathbf{F}})\mathbf{K}^*. \tag{6.13}$$

Problem 6.1. Assume $\mathbf{A}$ and $\mathbf{B}$ to be invertible and show that $(\mathbf{AB})^* = \mathbf{A}^*\mathbf{B}^*$.

Recall the Legendre–Hadamard condition (2.61):

$$\mathbf{a} \otimes \mathbf{n} \cdot \Psi_{\mathbf{FF}}[\mathbf{a} \otimes \mathbf{n}] \geq 0 \quad \text{for all} \quad \mathbf{a} \otimes \mathbf{n}. \tag{6.14}$$

To interpret this in terms of W, we differentiate (6.13), at fixed $\mathbf{K}$, on a one-parameter family $\mathbf{H}(u) = \mathbf{F}(u)\mathbf{K}$, reaching

$$W_{\mathbf{HH}}[\dot{\mathbf{F}}\,\mathbf{K}] = (\Psi_{\mathbf{FF}}[\dot{\mathbf{F}}])\mathbf{K}^*, \tag{6.15}$$

where the superposed dot is the derivative with respect to u, evaluated at $u = 0$, say. Scalar multiplication by $\dot{\mathbf{F}}\,\mathbf{K}$ and use of the identity $\mathbf{A} \cdot \mathbf{BC} = \mathbf{B}^t\mathbf{A} \cdot \mathbf{C}$, with $\mathbf{B} = \dot{\mathbf{F}}$ and $\mathbf{C} = \mathbf{K}$, yields

$$\dot{\mathbf{F}}\mathbf{K} \cdot W_{\mathbf{HH}}[\dot{\mathbf{F}}\,\mathbf{K}] = \mathcal{J}_K \dot{\mathbf{F}} \cdot \Psi_{\mathbf{FF}}[\dot{\mathbf{F}}], \tag{6.16}$$

which we apply with $\dot{\mathbf{F}} = \mathbf{a} \otimes \mathbf{n}$ to express (6.14) in the form

$$\mathbf{a} \otimes \mathbf{m} \cdot W_{\mathbf{HH}}(\mathbf{H})[\mathbf{a} \otimes \mathbf{m}] \geq 0 \quad \text{for all} \quad \mathbf{a} \otimes \mathbf{m}, \tag{6.17}$$

with $\mathbf{m} = \mathbf{K}^t\mathbf{n}$. It follows that Ψ satisfies the Legendre–Hadamard condition at $\mathbf{F}$ if and only if W satisfies it at $\mathbf{H} = \mathbf{FK}$.

Proceeding exactly as in the derivation of (2.73), we have

$$W_{\mathbf{HH}}(\mathbf{H})[\mathbf{A}] = \mathbf{A}\hat{\mathbf{S}} + \mathbf{H}U_{\hat{\mathbf{E}}\hat{\mathbf{E}}}(\hat{\mathbf{E}})[\mathbf{H}^t\mathbf{A}] \tag{6.18}$$

for any tensor $\mathbf{A}$.

We impose the normalization condition $W(\mathbf{1}) = 0$, where $\mathbf{1}$ is the shifter discussed in Section 2.3. Equation (6.7) then implies that $W(\mathbf{H}) = 0$ if $\mathbf{H} \in Orth^+$. We assume the converse to also be true, and hence that $W(\mathbf{H}) = 0$ *if and only if* $\mathbf{H} \in Orth^+$. In view of our assumption that $\kappa_i(p)$ is stress free, we further assume that $\hat{\mathbf{S}} = \mathbf{0}$ if and only if $\hat{\mathbf{E}} = \mathbf{0}$; i.e., if and only if $\mathbf{H} \in Orth^+$, and, as in Section 2.3, that the classical fourth-order tensor of elastic moduli,

$$\mathcal{C} = U_{\hat{\mathbf{E}}\hat{\mathbf{E}}|\hat{\mathbf{E}}=\mathbf{0}}, \tag{6.19}$$

is positive definite in the sense that $\mathbf{A} \cdot \mathcal{C}[\mathbf{A}] > 0$ for all non-zero symmetric $\mathbf{A}$. This is precisely the same tensor that we defined in (2.77). Our assumptions imply that

$$W_{\mathbf{HH}}(\mathbf{Q})[\mathbf{A}] = \mathbf{Q}\mathcal{C}[\mathbf{Q}^t\mathbf{A}] \tag{6.20}$$

for any rotation $\mathbf{Q}$, and hence, as shown in Section 2.3, that the Legendre–Hadamard condition is automatically satisfied at zero elastic strain.

The presumed smallness of $|\hat{\mathbf{E}}|$ justifies the low-order Taylor expansion

$$U(\hat{\mathbf{E}}) = U(\mathbf{0}) + \hat{\mathbf{E}} \cdot U_{\hat{\mathbf{E}}}(\mathbf{0}) + \tfrac{1}{2}\hat{\mathbf{E}} \cdot \mathcal{C}[\hat{\mathbf{E}}] + o(|\hat{\mathbf{E}}|^2) \tag{6.21}$$

of the strain-energy function. Our assumptions then imply that the strain energy is approximated at leading order by the homogeneous quadratic function

$$U \simeq \tfrac{1}{2}\hat{\mathbf{E}} \cdot \mathcal{C}[\hat{\mathbf{E}}], \tag{6.22}$$

yielding

$$\hat{\mathbf{S}} \simeq \mathcal{C}[\hat{\mathbf{E}}] \tag{6.23}$$

at leading order in the elastic strain. Because the norm $|\hat{\mathbf{E}}|$ of the elastic strain is invariably small in metals—recall the discussion in Section 1.1—this approximation suffices for essentially all situations of practical interest. Moreover, the positive definiteness of $\mathcal{C}$ ensures that (6.23) is invertible, and hence that

$$\hat{\mathbf{E}} \simeq \mathcal{L}[\hat{\mathbf{S}}], \tag{6.24}$$

where $\mathcal{L} = \mathcal{C}^{-1}$ is the fourth-order compliance tensor. The modulus and compliance tensors possess major symmetry and both minor symmetries, as discussed in Section 2.3.

Writing (6.22) in the form $U \simeq \frac{1}{2}\hat{\mathbf{S}} \cdot \hat{\mathbf{E}}$, we then have

$$U \simeq \tfrac{1}{2}\hat{\mathbf{S}} \cdot \mathcal{L}[\hat{\mathbf{S}}]. \tag{6.25}$$

Our assumptions thus imply that the energy is approximated by a positive definite function of $\hat{\mathbf{E}}$ or $\hat{\mathbf{S}}$.

Recall that the solid is undistorted in $\kappa_i(p)$; therefore,

$$W(\mathbf{H}) = W(\mathbf{HR}), \tag{6.26}$$

where $\mathbf{R} \in g_{\kappa_i(p)}$ is a rotation, and hence

$$\begin{aligned} U(\hat{\mathbf{E}}) &= U(\tfrac{1}{2}(\mathbf{H}^t\mathbf{H}\text{-}\mathbf{I})) = U(\tfrac{1}{2}(\mathbf{R}^t\mathbf{H}^t\mathbf{H}\mathbf{R}\text{-}\mathbf{I})) \\ &= U(\tfrac{1}{2}\mathbf{R}^t(\mathbf{H}^t\mathbf{H}\text{-}\mathbf{I})\mathbf{R}) = U(\mathbf{R}^t\hat{\mathbf{E}}\mathbf{R}). \end{aligned} \tag{6.27}$$

This implies that the transformation $\hat{\mathbf{E}} \to \mathbf{R}^t\hat{\mathbf{E}}\mathbf{R}$ induces the transformation

$$\hat{\mathbf{S}} \to \mathbf{R}^t\hat{\mathbf{S}}\mathbf{R}. \tag{6.28}$$

To see this consider a one-parameter family $\hat{\mathbf{E}}(u)$ of elastic strains, with $u \in (-u_0, u_0)$. Following a procedure detailed in the book by Gurtin et al., we define

$$f(u) = U(\hat{\mathbf{E}}(u)). \tag{6.29}$$

Then,

$$\dot{f} = U_{\hat{\mathbf{E}}}(\hat{\mathbf{E}}_0) \cdot \mathbf{A}, \tag{6.30}$$

where $\dot{f} = f'(u)_{|u=0}$, $\hat{\mathbf{E}}_0 = \hat{\mathbf{E}}(0)$ and $\mathbf{A} = \hat{\mathbf{E}}'(u)_{|u=0}$. According to (6.27) it is also true that

$$f(u) = U(\mathbf{R}^t\hat{\mathbf{E}}(u)\mathbf{R}), \tag{6.31}$$

and hence that

$$\dot{f} = U_{\hat{\mathbf{E}}}(\mathbf{R}^t\hat{\mathbf{E}}_0\mathbf{R}) \cdot \mathbf{R}^t\mathbf{A}\mathbf{R}. \tag{6.32}$$

Problem 6.2. Prove the identity $\mathbf{A} \cdot \mathbf{BCD} = \mathbf{B}^t\mathbf{A}\mathbf{D}^t \cdot \mathbf{C}$.

Comparing (6.30) and (6.32) gives

$$[\mathbf{R}U_{\hat{\mathbf{E}}}(\mathbf{R}^t\hat{\mathbf{E}}_0\mathbf{R})\mathbf{R}^t - U_{\hat{\mathbf{E}}}(\hat{\mathbf{E}}_0)] \cdot \mathbf{A} = 0, \tag{6.33}$$

implying, as the term in braces is symmetric, $\mathbf{A}$ is an arbitrary symmetric tensor, $\mathbf{R}$ is orthogonal, and dropping the subscript $_0$, that

$$U_{\hat{\mathbf{E}}}(\mathbf{R}^t\hat{\mathbf{E}}\mathbf{R}) = \mathbf{R}^t U_{\hat{\mathbf{E}}}(\hat{\mathbf{E}})\mathbf{R}, \tag{6.34}$$

which is just (6.28). That is, the stress evaluated at $\mathbf{R}^t\hat{\mathbf{E}}\mathbf{R}$ equals the stress evaluated at $\hat{\mathbf{E}}$, pre-multiplied by $\mathbf{R}^t$ and post-multiplied by $\mathbf{R}$.

Problem 6.3. The classical stress–strain relation for isotropic materials is

$$\hat{\mathbf{S}} = \lambda(tr\hat{\mathbf{E}})\mathbf{I} + 2\mu\hat{\mathbf{E}},$$

where λ and μ, the *Lamé moduli*, are constants (see the Appendix to this section). Verify (6.28) for arbitrary orthogonal $\mathbf{R}$.

Note that we have extended $\mathbf{R}$ from the group of rotations to the group of orthogonal tensors. This is permissible because, if $\mathbf{R} \in Orth^+$ satisfies (6.27), then $-\mathbf{R} \in Orth$ does too. Because the group *Orth* is obtained by adding $-\mathbf{I}$ to the elements of the group $Orth^+$, it follows that the symmetry group is effectively a subgroup of *Orth*. In the case of isotropy it coincides with *Orth*. For example, if the 180° rotation $\mathbf{R} = 2\mathbf{e} \otimes \mathbf{e} - \mathbf{I}$ about a unit vector $\mathbf{e}$ belongs to the symmetry group, then so does the reflection $-\mathbf{R} = \mathbf{I} - 2\mathbf{e} \otimes \mathbf{e}$ through the plane with normal $\mathbf{e}$, and vice versa. In this way we accommodate the reflection symmetries observed in crystals, a kind of symmetry that cannot be associated with an orientation-preserving transformation having $\mathcal{J}_R > 0$.

Appendix: Isotropy

The strain-energy function $U(\hat{\mathbf{E}})$ for isotropic materials is such that $U(\hat{\mathbf{E}}) = U(\mathbf{R}^t\hat{\mathbf{E}}\mathbf{R})$ for any orthogonal $\mathbf{R}$. It therefore depends on $\hat{\mathbf{E}}$ via the scalars $tr\hat{\mathbf{E}}$, $tr(\hat{\mathbf{E}}^2)$, and $tr(\hat{\mathbf{E}}^3)$. Sufficiency is immediate because these scalars are invariant under $\hat{\mathbf{E}} \to \mathbf{R}^t\hat{\mathbf{E}}\mathbf{R}$ for every orthogonal $\mathbf{R}$. To prove necessity, we first note that the principal invariants of $\hat{\mathbf{E}}$ are $I_1 = tr\hat{\mathbf{E}}$, $I_2 = \frac{1}{2}[(tr\hat{\mathbf{E}})^2 - tr(\hat{\mathbf{E}}^2)]$, and $I_3 = \det\hat{\mathbf{E}}$; these are the coefficients in the cubic characteristic equation for the eigenvalues of $\hat{\mathbf{E}}$, and also in the Cayley–Hamilton formula for $\hat{\mathbf{E}}$. See, for example, the book by Gurtin et al. The latter may be used to express $\det\hat{\mathbf{E}}$ in terms of the list $\{tr\hat{\mathbf{E}}, tr(\hat{\mathbf{E}}^2), tr(\hat{\mathbf{E}}^3)\}$, and so this list determines the principal invariants.

We need to show that if $I_k(\mathbf{A}) = I_k(\mathbf{B})$; $k = 1, 2, 3$ for any symmetric tensors $\mathbf{A}$,$\mathbf{B}$, then $U(\mathbf{A}) = U(\mathbf{B})$. For $U(\hat{\mathbf{E}})$ is then determined by the $I_k(\hat{\mathbf{E}})$, and hence by $\{tr\hat{\mathbf{E}}, tr(\hat{\mathbf{E}}^2), tr(\hat{\mathbf{E}}^3)\}$. Thus, suppose $I_k(\mathbf{A}) = I_k(\mathbf{B})$; $k = 1, 2, 3$. The characteristic equation for eigenvalues then implies that **A**,**B** have the same (real) eigenvalues λ_i. By the spectral representation for symmetric tensors, we have

$$\mathbf{A} = \sum \lambda_i \mathbf{a}_i \otimes \mathbf{a}_i \quad \text{and} \quad \mathbf{B} = \sum \lambda_i \mathbf{b}_i \otimes \mathbf{b}_i,$$

where $\{\mathbf{a}_i\}$ and $\{\mathbf{b}_i\}$ are orthornormal sets. Then $\mathbf{Q} = \mathbf{a}_i \otimes \mathbf{b}_i$ is orthogonal, and $\mathbf{b}_i = \mathbf{Q}^t \mathbf{a}_i$, yielding $\mathbf{B} = \mathbf{Q}^t\mathbf{A}\mathbf{Q}$ and $U(\mathbf{B}) = U(\mathbf{Q}^t\mathbf{A}\mathbf{Q})$. But $U(\mathbf{Q}^t\mathbf{A}\mathbf{Q}) = U(\mathbf{A})$ by hypothesis; thus, $U(\mathbf{A}) = U(\mathbf{B})$, and necessity is proved.

In the case of isotropy the most general homogeneous quadratic strain-energy function is thus of the form $U(\hat{\mathbf{E}}) = \frac{1}{2}\lambda(tr\hat{\mathbf{E}})^2 + \mu tr(\hat{\mathbf{E}}^2)$, where λ and μ are constants, and, as is well known, for this to be positive definite it is necessary and sufficient that $3\lambda + 2\mu > 0$ and $\mu > 0$. To obtain the stress we use

$$U_{\hat{\mathbf{E}}} \cdot \hat{\mathbf{E}}' = \lambda(tr\hat{\mathbf{E}})\mathbf{I} \cdot \hat{\mathbf{E}}' + \mu(\hat{\mathbf{E}} \cdot \hat{\mathbf{E}})' = [\lambda(tr\hat{\mathbf{E}})\mathbf{I} + 2\mu\hat{\mathbf{E}}] \cdot \hat{\mathbf{E}}',$$

where the primes are the derivatives, with respect to the parameter, of a one-parameter family of strains, and thus conclude, from (2.33), that $\hat{\mathbf{S}}$ is as stated in Problem 6.3.

6.2 Surface dislocations and stress relaxation

We have seen that if the plastic deformation tensor $\mathbf{K}^{-1}$ is a smooth function of $\mathbf{x}$ in κ, then there exists a dislocation density $\boldsymbol{\alpha}_\mathcal{K} = Curl\mathbf{K}^{-1}$ such that

$$\mathbf{B}(\Omega_\perp, t) = \int_{\Omega_\perp} \boldsymbol{\alpha}_\mathcal{K}^t \boldsymbol{\nu}_\perp dA, \tag{6.35}$$

where $\Omega_\perp \subset \kappa$ is any simply connected surface with unit-normal field $\boldsymbol{\nu}_\perp$, with $\mathbf{B}(\Omega_\perp, t)$ the associated Burgers vector. Consider a surface Ω across which $\mathbf{K}^{-1}$ is discontinuous, and suppose $\Omega_\perp$ intersects Ω orthogonally. Let Γ be the curve of intersection. If $\mathbf{K}^{-1}$ is smooth in the regions on either side of Ω, then Stokes' theorem (3.120) may be applied to the individual parts $\Omega_\perp^\pm$ of $\Omega_\perp$ separated by Ω. Adding the two expressions and using $\Gamma = \partial\Omega_\perp^+ \cap \partial\Omega_\perp^-$, we then have

$$\int_{\Omega_\perp} \boldsymbol{\alpha}_\mathcal{K}^t \boldsymbol{\nu}_\perp dA = \int_{\Omega_\perp} (Curl\mathbf{K}^{-1})^t \boldsymbol{\nu}_\perp dA = \int_{\partial\Omega_\perp} \mathbf{K}^{-1} d\mathbf{x} + \int_\Gamma [\mathbf{K}^{-1}] d\mathbf{x}, \tag{6.36}$$

where

$$[\mathbf{K}^{-1}] = (\mathbf{K}^{-1})_+ - (\mathbf{K}^{-1})_- \tag{6.37}$$

is the discontinuity of $\mathbf{K}^{-1}$ on Ω. Here we use subscripts $\pm$, respectively, to denote limits as Ω, with unit-normal field $\boldsymbol{\nu}$, is approached from the regions into which $\boldsymbol{\nu}$ and $-\boldsymbol{\nu}$ are directed; and, $d\mathbf{x} = \boldsymbol{\nu} \times \mathbf{T} dS$, where $\mathbf{T} = \boldsymbol{\nu}_{\perp|\Gamma}$ and S measures arclength on Γ. Note that

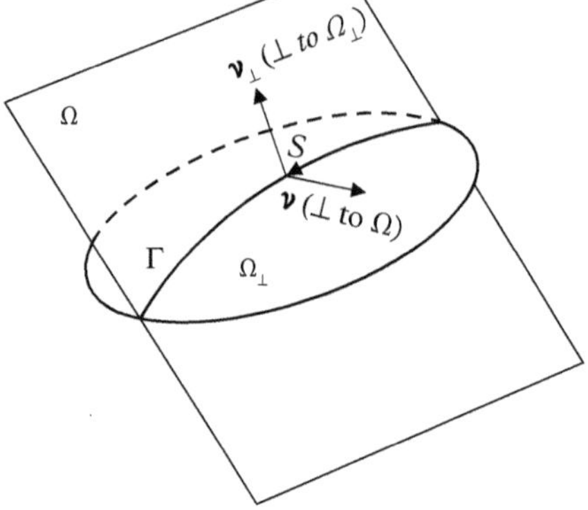

Figure 6.1 *A surface Ω across which* **K** *is discontinuous, intersecting a material suface $\boldsymbol{\Omega}_\perp$ orthogonally along the curve Γ.*

$\mathbf{T} \in T_{\Omega(\mathbf{x},t)}$, the tangent plane to Ω at the point $\mathbf{x}$ at time t (Figure 6.1), and that $\boldsymbol{\nu} \times \mathbf{T}$ is tangential to Γ as $\partial\Omega_\perp^+$ is traversed in the sense of Stokes' theorem. Thus, on Γ we have $[\mathbf{K}^{-1}]d\mathbf{x} = [\mathbf{K}^{-1}](\boldsymbol{\nu} \times \mathbf{T})dS$.

For a given surface $\boldsymbol{\Omega}$, with unit-normal field $\boldsymbol{\nu}$, the foregoing construction allows for infinitely many surfaces $\boldsymbol{\Omega}_\perp$ with corresponding intersection curves Γ and unit-normal fields $\boldsymbol{\nu}_{\perp|\Gamma}$. Thus, $\mathbf{T}$ is an arbitrary unit tangent to $T_{\Omega(\mathbf{x},t)}$, and $[\mathbf{K}^{-1}](\boldsymbol{\nu} \times \mathbf{T})$ may be regarded as a linear function of the variable $\mathbf{T} \in T_{\Omega(\mathbf{x},t)}$. Accordingly, there is a tensor field $\beta_{\mathcal{K}}$ defined on Ω—the *referential surface dislocation density*—such that

$$[\mathbf{K}^{-1}](\mathbf{T} \times \boldsymbol{\nu}) = \beta_{\mathcal{K}}^t \mathbf{T}. \tag{6.38}$$

Because the action of $\beta_{\mathcal{K}}^t$ on $\boldsymbol{\nu}$ is undefined in this expression, we can extend its domain to three-space and impose $\beta_{\mathcal{K}}^t \boldsymbol{\nu} = \mathbf{0}$ without loss of generality. From (6.36) and noting that $\mathbf{T} \times \boldsymbol{\nu} = -\boldsymbol{\nu} \times \mathbf{T}$, we have

$$\mathbf{B}(\Omega_\perp, t) = \int_{\partial\Omega_\perp} \mathbf{K}^{-1} d\mathbf{x} = \int_{\Omega_\perp} \alpha_{\mathcal{K}}^t \boldsymbol{\nu}_\perp dA + \int_\Gamma \beta_{\mathcal{K}}^t \mathbf{T}\, dS. \tag{6.39}$$

Equation (6.38) can be solved for $\beta_{\mathcal{K}}^t$ by introducing an orthonormal basis $\{\mathbf{T}_1, \mathbf{T}_2\}$ for $T_{\Omega(\mathbf{x},t)}$ and writing the three-dimensional identity as

$$\mathbf{I} = \boldsymbol{\nu} \otimes \boldsymbol{\nu} + \mathbf{T}_\alpha \otimes \mathbf{T}_\alpha \tag{6.40}$$

in which α is summed from one to two. Then,

$$[\mathbf{K}^{-1}] = [\mathbf{K}^{-1}]\mathbf{I} = \mathbf{k} \otimes \boldsymbol{\nu} + [\mathbf{K}^{-1}]\mathbf{T}_\alpha \otimes \mathbf{T}_\alpha, \quad \text{where} \quad \mathbf{k} = [\mathbf{K}^{-1}]\boldsymbol{\nu}. \tag{6.41}$$

Further, if we choose $\mathbf{T}_\alpha$ such that $\{\mathbf{T}_1, \mathbf{T}_2, \boldsymbol{\nu}\}$ is positively oriented, then (6.38) furnishes

$$[\mathbf{K}^{-1}]\mathbf{T}_1 = [\mathbf{K}^{-1}](\mathbf{T}_2 \times \boldsymbol{\nu}) = \boldsymbol{\beta}^t_{\mathcal{K}}\mathbf{T}_2, \tag{6.42}$$

and, similarly,

$$[\mathbf{K}^{-1}]\mathbf{T}_2 = -[\mathbf{K}^{-1}](\mathbf{T}_1 \times \boldsymbol{\nu}) = -\boldsymbol{\beta}^t_{\mathcal{K}}\mathbf{T}_1. \tag{6.43}$$

Thus,

$$[\mathbf{K}^{-1}] = \mathbf{k} \otimes \boldsymbol{\nu} - \beta^t_{\mathcal{K}}\boldsymbol{\epsilon}_{(\boldsymbol{\nu})}, \tag{6.44}$$

where

$$\boldsymbol{\epsilon}_{(\boldsymbol{\nu})} = \mathbf{T}_1 \otimes \mathbf{T}_2 - \mathbf{T}_2 \otimes \mathbf{T}_1 \tag{6.45}$$

is the two-dimensional permutation tensor on $T_{\Omega(\mathrm{x},t)}$. The name is justified by the fact that it is independent of the particular choice of orthonormal vectors $\mathbf{T}_\alpha$ in $T_{\Omega(\mathrm{x},t)}$ and thus has an invariant meaning.

Problem 6.4. Prove this.

To solve (6.44) we post-multiply both sides by $\boldsymbol{\epsilon}_{(\boldsymbol{\nu})}$ and use $(\mathbf{k} \otimes \boldsymbol{\nu})\boldsymbol{\epsilon}_{(\boldsymbol{\nu})} = \mathbf{k} \otimes \boldsymbol{\epsilon}^t_{(\boldsymbol{\nu})}\boldsymbol{\nu}$ with $\boldsymbol{\epsilon}^t_{(\boldsymbol{\nu})}\boldsymbol{\nu} = \mathbf{0}$, together with $\boldsymbol{\epsilon}^2_{(\boldsymbol{\nu})} = -\mathbf{T}_\alpha \otimes \mathbf{T}_\alpha = \boldsymbol{\nu} \otimes \boldsymbol{\nu} - \mathbf{I}$, finally reaching

$$\beta^t_{\mathcal{K}} = [\mathbf{K}^{-1}]\boldsymbol{\epsilon}_{(\boldsymbol{\nu})}. \tag{6.46}$$

As might be expected we can repeat the entire exercise with a simply connected surface ω in κ_t and a surface $\omega_\perp$ that intersects it orthogonally. We would then establish the representation

$$\mathbf{b}(\omega_\perp, t) = \int_{\omega_\perp} \alpha^t_{\mathcal{K}_t}\mathbf{n}_\perp da + \int_\gamma \beta^t_{\mathcal{K}_t}\mathbf{t} ds \tag{6.47}$$

for the relevant Burgers vector, where $\gamma = \partial\omega^+_\perp \cap \partial\omega^-_\perp$; s is arclength on γ; $\mathbf{t} \in T_{\omega(\mathrm{y},t)}$, with unit normal $\mathbf{n}$; $\mathbf{n} \times \mathbf{t}$ is the unit tangent to γ; and $\beta_{\mathcal{K}_t}$ —the *spatial surface dislocation density*—is given by

$$\beta^t_{\mathcal{K}_t} = [\mathbf{H}^{-1}]\boldsymbol{\epsilon}_{(\mathbf{n})}, \tag{6.48}$$

in which $[\mathbf{H}^{-1}]$ is the jump of $\mathbf{H}^{-1}$ across ω and $\boldsymbol{\epsilon}_{(\mathbf{n})}$ is the two-dimensional permutation tensor on $T_{\omega(\mathbf{y},t)}$. The counterpart of (6.44) is

$$[\mathbf{H}^{-1}] = \mathbf{h} \otimes \mathbf{n} - \beta^t_{\kappa_t} \boldsymbol{\epsilon}_{(\mathbf{n})}, \quad \text{where} \quad \mathbf{h} = [\mathbf{H}^{-1}]\mathbf{n}. \tag{6.49}$$

Note that in the foregoing discussion there is no requirement that Ω be a material surface. That is, it may evolve relative to the material and thus be associated with a moving surface in the reference configuration κ. Therefore, the notion of surface dislocation applies not only to grain boundaries but also to evolving phase interfaces, to be discussed in Section 6.7.

Problem 6.5. If Ω is a material surface, then the referential and spatial surface dislocation densities are not independent. Derive the relationship between them, and show that if one vanishes, then both vanish.

It is instructive to revisit the foregoing procedure with $\mathbf{K}^{-1}$ replaced by the deformation gradient $\mathbf{F} = \nabla\chi$. Assuming χ to be a continuous function of $\mathbf{x}$, and that its gradient suffers a jump across Ω, we have

$$0 = \int_{\partial\Omega_\perp} d\chi = \int_{\partial\Omega_\perp} (\nabla\chi)d\mathbf{x} = \int_{\Omega_\perp} (Curl\mathbf{F})^t \boldsymbol{\nu}_\perp dA + \int_\Gamma [\mathbf{F}]d\mathbf{x}. \tag{6.50}$$

Problem 6.6. Show that $Curl\mathbf{F}$ vanishes in the regions separated by the surface Ω, assuming χ to be a twice-differentiable function of $\mathbf{x}$ therein.

Using $d\mathbf{x} = \hat{\mathbf{T}}dS$ on Γ, where $\hat{\mathbf{T}} = \boldsymbol{\nu} \times \mathbf{T}$ is the unit tangent, it follows that

$$\int_\Gamma [\mathbf{F}]\hat{\mathbf{T}}dS = \mathbf{0}. \tag{6.51}$$

As $\Omega_\perp$ is arbitrary, and hence so too $\Gamma = \Omega \cap \Omega_\perp$, and as $\hat{\mathbf{T}} \cdot \boldsymbol{\nu} = 0$, we may localize and conclude that $[\mathbf{F}]\hat{\mathbf{T}} = \mathbf{0}$ for all $\hat{\mathbf{T}} \in T_{\Omega(\mathbf{x},t)}$. Then, with

$$[\mathbf{F}] = [\mathbf{F}]\mathbf{I} = [\mathbf{F}]\boldsymbol{\nu} \otimes \boldsymbol{\nu} + [\mathbf{F}]\mathbf{T}_\alpha \otimes \mathbf{T}_\alpha, \tag{6.52}$$

where $\{\mathbf{T}_\alpha\}$ is any orthonormal basis for $T_{\Omega(\mathbf{x},t)}$, with unit normal $\boldsymbol{\nu}$, we arrive at *Hadamard's lemma*

$$[\mathbf{F}] = \mathbf{f} \otimes \boldsymbol{\nu} \quad \text{on} \quad \Omega, \tag{6.53}$$

where $\mathbf{f} = [\mathbf{F}]\boldsymbol{\nu}$. See Chapter C in the treatise by Truesdell and Toupin for a comprehensive discussion of discontinuity relations of this kind. We conclude that $[\mathbf{F}]$ is a *rank-one* tensor at a *coherent* interface Ω; that is, at an interface across which χ is continuous but $\mathbf{F}$ may be discontinuous. This is a severe constraint on the limiting values $\mathbf{F}_{\pm}$ at the surface Ω.

Here, the rank of a tensor is the dimension of its image space:

$$Rank\mathbf{A} = \dim(Span\{\mathbf{Av}\}) \tag{6.54}$$

for any three-vector $\mathbf{v}$. In our example, $[\mathbf{F}]\mathbf{v} = (\boldsymbol{\nu} \cdot \mathbf{v})\mathbf{f}$ and $Span\{[\mathbf{F}]\mathbf{v}\} = Span\mathbf{f}$ is one-dimensional.

In contrast, (6.49) shows that $Rank[\mathbf{H}^{-1}] = 3$, and hence that $[\mathbf{H}^{-1}]$ is a full-rank tensor if surface dislocation is present. In this case there is no constraint on the limiting values $\mathbf{H}_{\pm}$. To better understand the significance of this result, consider adjoining stress-free crystal grains separated by the interface Ω. Because the associated modulus tensors are positive definite, it follows that they are also free of elastic strain and their respective strain energies vanish. In particular, $\mathbf{H}_{\pm} \in Orth^{+}$ and (6.49) yields a (generally non-zero) surface dislocation density $\beta_{\mathcal{K}_t}$. The so-called *tilt* and *twist* boundaries furnish illustrative examples—see Section 4.4 in Lardner's book and Section 8.2.2.3 in the book by Nabarro. However, if $\beta_{\mathcal{K}_t}$ should happen to vanish, then from (6.49) we would have $(\mathbf{H}^{-1})_{+} = (\mathbf{H}^{-1})_{-} + \mathbf{h} \otimes \mathbf{n}$. If one of the adjoining grains is stress free, so that $\mathbf{H}_{-}$, say, is a rotation, then $\mathbf{H}_{+}$ is not a rotation and the adjacent grain is necessarily stressed. Because of the positive definiteness of (6.22), its energy is also positive and hence so too the total energy of the two grains together. Accordingly, surface dislocation induced by plastic deformation—see (6.46) and Problem 6.5—is an additional interfacial degree of freedom which is available to minimize the energy in adjacent grains. Presumably this is related to the fact that metallic crystals most often occur as aggregates of adjoining grains rather than singly.

6.3 Dissipation due to plastic evolution

A central feature of plastic evolution is its irreversibility. This is intimately connected with the notion of energy dissipation. To introduce the concept of dissipation attending plastic evolution in as simple a manner as possible, we confine attention to smooth deformations in the present section. Deformations with discontinuous gradients will be taken up in Section 6.7.

We begin with the energy balance (1.44) and the strain energy

$$\mathcal{U}(\pi, t) = \int_{\pi} \Psi(\mathbf{F},\mathbf{K})dV; \quad \text{where} \quad \Psi(\mathbf{F},\mathbf{K}) = \mathcal{J}_K^{-1} W(\mathbf{H}) \quad \text{with} \quad \mathbf{H} = \mathbf{FK}. \tag{6.55}$$

The total mechanical energy is $\mathcal{U}(\pi, t) + \mathcal{K}(\pi, t)$, where $\mathcal{K}$ is the kinetic energy, and we define the dissipation $\mathcal{D}$, as in Problem 2.6, by

$$\mathcal{D}(\pi, t) = \mathcal{P}(\pi, t) - \tfrac{d}{dt}[\mathcal{U}(\pi, t) + \mathcal{K}(\pi, t)], \tag{6.56}$$

where $\mathcal{P}(\pi, t)$ is the power supplied to the arbitrary part π of κ. We suppose that

$$\mathcal{D}(\pi, t) \geq 0 \quad \text{for all} \quad \pi \subset \kappa, \tag{6.57}$$

this effectively serving as a surrogate for the second law of thermodynamics in the present, purely mechanical, setting.

It follows from (1.44) that $\mathcal{D} = \mathcal{S} - \frac{d}{dt}\mathcal{U}$, where $\mathcal{S}$ is the stress power. Combining with (2.34), it follows immediately that

$$\mathcal{D}(\pi, t) = \int_\pi D dV, \tag{6.58}$$

where

$$D = \mathbf{P} \cdot \dot{\mathbf{F}} - \dot{\Psi}, \tag{6.59}$$

and localizing (6.57) yields

$$D \geq 0 \quad \text{at all} \quad \mathbf{x} \in \kappa. \tag{6.60}$$

To obtain a useful expression for the dissipation we differentiate $(6.55)_2$:

$$\dot{\Psi} = \mathcal{J}_K^{-1}[\dot{W} - (\dot{\mathcal{J}}_K/\mathcal{J}_K)W]. \tag{6.61}$$

Here we use the identity $\dot{\mathcal{J}}_K/\mathcal{J}_K = \mathbf{K}^{-t} \cdot \dot{\mathbf{K}}$ together with

$$\dot{W} = W_{\mathbf{H}} \cdot \dot{\mathbf{H}} = W_{\mathbf{H}} \cdot (\dot{\mathbf{F}}\mathbf{K} + \mathbf{F}\dot{\mathbf{K}}), \tag{6.62}$$

which is reduced, with the aid of the identities $\mathbf{A} \cdot \mathbf{BC} = \mathbf{AC}^t \cdot \mathbf{B} = \mathbf{B}^t\mathbf{A} \cdot \mathbf{C}$, to

$$\dot{W} = W_{\mathbf{H}}\mathbf{K}^t \cdot \dot{\mathbf{F}} + \mathbf{F}^t W_{\mathbf{H}} \cdot \dot{\mathbf{K}}. \tag{6.63}$$

Recalling (6.13), which yields $W_{\mathbf{H}}\mathbf{K}^t = \mathcal{J}_K\mathbf{P}$ and $\mathbf{F}^t W_{\mathbf{H}} = \mathcal{J}_K\mathbf{F}^t\mathbf{P}\mathbf{K}^{-t}$, Eq. (6.61) is finally reduced to

$$\dot{\Psi} = \mathbf{P} \cdot \dot{\mathbf{F}} - \mathbb{E} \cdot \dot{\mathbf{K}}\mathbf{K}^{-1}, \tag{6.64}$$

where

$$\mathbb{E} = \Psi\mathbf{I} - \mathbf{F}^t\mathbf{P} \tag{6.65}$$

is the *Eshelby tensor*. Accordingly, the local dissipation is given in terms of the rate of plastic deformation $\dot{\mathbf{K}}$ by

$$D = \mathbb{E} \cdot \dot{\mathbf{K}}\mathbf{K}^{-1}. \tag{6.66}$$

This important result, first announced by Epstein and Maugin, highlights the role of the Eshelby tensor as the driving force for dissipation. We will use it in Section 6.5, in conjunction with inequality (6.60), to derive restrictions on constitutive equations for the plastic evolution $\dot{\mathbf{K}}$. Observe, from (6.64) and the chain rule, that

$$\Psi_{\mathbf{F}}(\mathbf{F},\mathbf{K}) = \mathbf{P} \quad \text{and} \quad \Psi_{\mathbf{K}}(\mathbf{F},\mathbf{K}) = -\mathbb{E}\mathbf{K}^{-t}, \tag{6.67}$$

the first of these being simply a restatement of (6.12).

The expression (6.66) for D makes clear the fact that the dissipation vanishes in the absence of plastic evolution; i.e., $D=0$ if $\dot{\mathbf{K}} = \mathbf{0}$. On the basis of empirical observation, we introduce the hypothesis that plastic evolution is inherently dissipative; thus, we suppose that $D \neq 0$ *if and only if* $\dot{\mathbf{K}} \neq \mathbf{0}$. In view of (6.60), this means that

$$\dot{\mathbf{K}} \neq \mathbf{0} \quad \text{if and only if} \quad D > 0. \tag{6.68}$$

It may be observed, from the definition (6.65), that the Eshelby tensor is purely referential, mapping the translation space T_κ to itself. It proves convenient to introduce a version of the Eshelby tensor, $\mathbb{E}'$, that maps $\kappa_i(p)$ to itself. This is given by the relation

$$\mathbb{E} = \mathcal{J}_K^{-1}\mathbf{K}^{-t}\mathbb{E}'\mathbf{K}^t. \tag{6.69}$$

To establish this we may proceed from the observation that if $\mathbb{E}^*$ is the spatial Eshelby tensor, derived by taking the current configuration as reference, i.e., by taking κ to coincide with κ_t at a particular instant t, then

$$\mathbb{E}^* = \psi\mathbf{i} - \mathbf{T}, \quad \text{with} \quad \psi = \mathcal{J}_F^{-1}\Psi, \tag{6.70}$$

where $\mathbf{i}$ is the spatial identity, so that

$$\mathbb{E} = \mathcal{J}_F\mathbf{F}^t\mathbb{E}^*\mathbf{F}^{-t}. \tag{6.71}$$

Thus, $\mathbb{E}$ is the *pullback* of $\mathbb{E}^*$ from κ_t to κ. It follows that $\mathbb{E}$ is the pullback of $\mathbb{E}'$ from $\kappa_i(p)$ to $T_\kappa(p)$, and $\mathbb{E}'$ is the pullback of $\mathbb{E}^*$ from $T_{\kappa_t}(p)$ to $\kappa_i(p)$. In particular,

$$\mathbb{E}' = \mathcal{J}_H\mathbf{H}^t\mathbb{E}^*\mathbf{H}^{-t} = W\mathbf{I} - \mathbf{H}^t W_{\mathbf{H}}, \tag{6.72}$$

which implies that $\mathbb{E}'$ is determined solely by $\mathbf{H}$ and hence purely elastic in origin. This is equivalent to

$$\mathbb{E}' = U\mathbf{I} - \mathbf{H}^t\mathbf{H}\hat{\mathbf{S}}, \tag{6.73}$$

which further implies that $\mathbb{E}'$ remains invariant if $\mathbf{H}$ is replaced by $\bar{\mathbf{Q}}\mathbf{H}$ for any rotation $\bar{\mathbf{Q}}$.

In the practically important case of small elastic strain, (6.22) and (6.23) yield the estimate

$$\mathbb{E}' = -\hat{\mathbf{S}} + o(|\hat{\mathbf{E}}|). \tag{6.74}$$

Thus, the Eshelby tensor, referred to $\kappa_i(p)$, is given, to leading order and apart from sign, by the Piola–Kirchhoff stress referred to the same state.

From (6.69) and the identity $\mathbf{A} \cdot \mathbf{BC} = \mathbf{AC}^t \cdot \mathbf{B}$ we have that

$$\mathbb{E} \cdot \dot{\mathbf{K}}\mathbf{K}^{-1} = \mathbb{E}\mathbf{K}^{-t} \cdot \dot{\mathbf{K}} = \mathcal{J}_K^{-1}\mathbf{K}^{-t}\mathbb{E}' \cdot \dot{\mathbf{K}} = \mathcal{J}_K^{-1}\mathbb{E}' \cdot \mathbf{K}^{-1}\dot{\mathbf{K}}. \tag{6.75}$$

Thus,

$$\mathcal{J}_K D = \mathbb{E}' \cdot \mathbf{K}^{-1}\dot{\mathbf{K}}, \tag{6.76}$$

and the assumption of inherent dissipativity is equivalent to the statement:

$$\dot{\mathbf{K}} \neq \mathbf{0} \quad \text{if and only if} \quad \mathbb{E}' \cdot \mathbf{K}^{-1}\dot{\mathbf{K}} > 0. \tag{6.77}$$

It is interesting to observe that if $\hat{\mathbf{E}} = \mathbf{0}$, then, according to our constitutive hypotheses, $U = 0$ and $\hat{\mathbf{S}} = \mathbf{0}$. Equations (6.73) and (6.76) then give $D = 0$ and (6.77) implies that there can be no plastic evolution. That is, without stress, there can be no change in the plastic deformation. This comports with the observed phenomenology.

Problem 6.7. Show that the dissipation D is invariant under material symmetry transformations.

6.4 Superposed rigid-body motions

We know that the symmetry of the Cauchy stress $\mathbf{T}$ is equivalent to the statement $W(\mathbf{H}) = W(\bar{\mathbf{Q}}\mathbf{H})$ for all rotations $\bar{\mathbf{Q}}$. This follows from $W_{\mathbf{H}} = \mathbf{TH}^*$, implying that $W_{\mathbf{H}} \cdot \mathbf{\Omega H}$ vanishes for any skew tensor $\mathbf{\Omega}$. The result then follows exactly as in Section 2.1. Because the argument leading to this conclusion is purely local, the rotation $\bar{\mathbf{Q}}$ may conceivably vary from one material point to another. This stands in contrast to the spatially uniform rotation $\mathbf{Q}\ (t)$ associated with a superposed rigid-body motion; our notation is intended to distinguish these cases explicitly. In a superposed rigid-body motion, the deformation $\boldsymbol{\chi}(\mathbf{x}, t)$ is changed to $\mathbf{Q}(t)\boldsymbol{\chi}(\mathbf{x}, t) + \mathbf{c}(t)$, as in (2.25), and this implies that $\mathbf{F}(= \nabla\boldsymbol{\chi})$ changes to $\mathbf{Q}(t)\mathbf{F}$. This simple deduction does not apply to $\mathbf{H}$ or $\mathbf{G}$, however. We therefore encounter the question of how the elastic and plastic deformations transform under a superposed rigid-body motion.

To address this we must keep in mind the fact that, at this stage in our analysis, $\bar{\mathbf{Q}}$ bears no *a priori* relation to a superposed rigid-body motion. On the contrary, our purpose is to establish such a relation. This requires that we introduce additional hypotheses. A natural one is that the strain energy W should be invariant; that is,

$$W(\mathbf{H}^+) = W(\mathbf{H}), \tag{6.78}$$

where $\mathbf{H}^+$ is the value of the elastic deformation associated with the superposed rigid motion. Because of the symmetry of the Cauchy stress, (6.8) is applicable and implies that $W(\mathbf{H}) = \hat{W}(\mathbf{H}^t\mathbf{H})$, where $\hat{W}(\mathbf{H}^t\mathbf{H}) = U(\frac{1}{2}(\mathbf{H}^t\mathbf{H}\text{-}\mathbf{I}))$. Then, $\hat{W}((\mathbf{H}^+)^t\mathbf{H}^+) = \hat{W}(\mathbf{H}^t\mathbf{H})$, i.e.,

$$\hat{W}(\mathbf{H}^t\mathbf{Z}^t\mathbf{Z}\mathbf{H}) = \hat{W}(\mathbf{H}^t\mathbf{H}), \quad \text{where} \quad \mathbf{Z} = \mathbf{H}^+\mathbf{H}^{-1}. \tag{6.79}$$

This must hold for all admissible $\mathbf{H}$, i.e., for all $\mathbf{H}$ with $\mathcal{J}_H > 0$. It therefore holds with $\mathbf{H}=\mathbf{1}$, the shifter, giving

$$\hat{W}(\mathbf{1}^t\mathbf{Z}^t\mathbf{Z}\mathbf{1}) = \hat{W}(\mathbf{I}). \tag{6.80}$$

The constitutive hypotheses introduced in Section 6.1 imply that the right-hand side vanishes, and that $\hat{W}(\mathbf{1}^t\mathbf{Z}^t\mathbf{Z}\mathbf{1}) > 0$ if $\mathbf{1}^t\mathbf{Z}^t\mathbf{Z}\mathbf{1} \neq \mathbf{I}$. The two statements are reconciled only if $\mathbf{1}^t\mathbf{Z}^t\mathbf{Z}\mathbf{1} = \mathbf{I}$, and this in turn yields $\mathbf{Z}^t\mathbf{Z} = \mathbf{i}$. Because $\mathcal{J}_Z > 0$, from $(6.79)_2$, we then have $\mathbf{Z} \in Orth^+$. We label this as $\bar{\mathbf{Q}}$. Conversely, this ensures that $(6.79)_1$ is satisfied for any admissible $\mathbf{H}$. We conclude that

$$\mathbf{H}^+ = \bar{\mathbf{Q}}\mathbf{H}, \tag{6.81}$$

where $\bar{\mathbf{Q}}(x,t)$ is an arbitrary rotation field, and thus recover (6.7). Then, $\mathbf{H}^+ = \bar{\mathbf{Q}}\mathbf{F}\mathbf{K}$, where $\mathbf{H}^+ = \mathbf{F}^+\mathbf{K}^+ = \mathbf{Q}\mathbf{F}\mathbf{K}^+$ in which $\mathbf{K}^+$ is the inverse plastic deformation associated with the superposed rigid-body motion. Hence,

$$\bar{\mathbf{Q}}\mathbf{F}\mathbf{K} = \mathbf{Q}\mathbf{F}\mathbf{K}^+, \tag{6.82}$$

implying that

$$\mathcal{J}_{K^+} = \mathcal{J}_K. \tag{6.83}$$

We would like to use this result to arrive at some definite conclusion about the relationship between $\mathbf{K}^+$ and $\mathbf{K}$, but this requires a further hypothesis. A natural one is that the dissipation is insensitive to superposed rigid-body motions. As justification we note that in continuum mechanics both the energy balance and the entropy production inequality have this property, the latter specializing to the dissipation inequality in a purely mechanical setting. See the discussion in the book by Liu, for example. Thus, we impose $D^+ = D$, where

$$\mathcal{J}_{K^+}D^+ = (\mathbb{E}')^+ \cdot (\mathbf{K}^+)^{-1}\dot{\mathbf{K}}^+, \tag{6.84}$$

in which $(\mathbb{E}')^+$ is the Eshelby tensor based on $\mathbf{H}^+$. From (6.73) and (6.81) it follows that $(\mathbb{E}')^+ = \mathbb{E}'$ and hence that $\mathbb{E}'$ is invariant under superposed rigid-body motions. The presumed invariance of the dissipation, combined with (6.83), then leads to

$$\mathbb{E}' \cdot (\mathbf{K}^+)^{-1}\dot{\mathbf{K}}^+ = \mathbb{E}' \cdot \mathbf{K}^{-1}\dot{\mathbf{K}}. \tag{6.85}$$

To explore the implications we define $\mathbf{Y} = \mathbf{K}^+\mathbf{K}^{-1}$ and note, from (6.83), that $\mathcal{J}_Y = 1$. Suppose $\mathbf{Y}(\mathbf{x}, t_0) = \mathbf{I}$, so that the superposed rigid motion commences at time t_0. We thus recast (6.85) in the form

$$\mathbb{E}' \cdot \mathbf{K}^{-1}\mathbf{Y}^{-1}\dot{\mathbf{Y}}\mathbf{K} = 0. \tag{6.86}$$

Recall that $\mathbf{K}$ requires the specification of a reference configuration, and that (5.13) furnishes the relationship between the values of $\mathbf{K}$ associated with two reference configurations κ_1 and κ_2, say. Thus, $\mathbf{K}_2 = \mathbf{R}\mathbf{K}_1$, where $\mathbf{R}$ is the gradient of the map from κ_1 to κ_2. We exploit the substantial freedom in the choice of reference configuration and choose this map such that $\mathbf{R} = \mathbf{K}_1^{-1}$ *at* the material point p in question, *at* a particular instant in time. The fact that $\mathbf{R}$ is a gradient and independent of time, whereas $\mathbf{K}_1$ is time-dependent and not a gradient, is thereby rendered irrelevant. Thus, our choice yields $\mathbf{K}_2 = \mathbf{I}$ at the material point and time in question. Applying (6.86) with $\kappa = \kappa_2$, we then have

$$\mathbb{E}' \cdot \mathbf{Y}^{-1}\dot{\mathbf{Y}} = 0, \quad \text{where} \quad \mathbf{Y} = \mathbf{K}_2^+, \tag{6.87}$$

with the time derivative evaluated at the considered instant. From (6.77) we then have that $\dot{\mathbf{Y}} = \mathbf{0}$ at the instant in question. As this instant is arbitrary, $\dot{\mathbf{Y}}$ vanishes at all instants, yielding $\mathbf{Y}(\mathbf{x}_2, t) = \mathbf{Y}(\mathbf{x}_2, t_0) = \mathbf{I}$, where $\mathbf{x}_2$ is the position of p in κ_2. Then, $\mathbf{K}_2^+ = \mathbf{K}_2$ and the transformation formula yields $\mathbf{R}\mathbf{K}_1^+ = \mathbf{K}_2^+ = \mathbf{K}_2 = \mathbf{R}\mathbf{K}_1$, implying that $\mathbf{K}_1^+ = \mathbf{K}_1$ and hence that $\mathbf{K}^+ = \mathbf{K}$ for all choices of reference configuration. Combining this result with (6.82), we conclude that $\bar{\mathbf{Q}}(\mathbf{x}, t) = \mathbf{Q}(t)$. Finally, the transformation formulas for $\mathbf{F}$,$\mathbf{H}$, and $\mathbf{K}$ under superposed rigid-body motions are

$$\mathbf{F} \to \mathbf{QF}, \quad \mathbf{H} \to \mathbf{QH}, \quad \text{and} \quad \mathbf{K} \to \mathbf{K}. \tag{6.88}$$

6.5 Yielding and plastic flow

In this section we extend the concept of yield, introduced in Section 1.1 in a one-dimensional setting, to three dimensions. Thus, we assume the onset of yield, and hence the possibility that $\dot{\mathbf{K}} \neq \mathbf{0}$, occurs when the elastic strain is such that $G(\hat{\mathbf{E}}) = 0$, where G is an appropriate threshold function, or *yield function*, pertaining to the material at hand. In a crystalline material, for example, this implies that yield occurs when the lattice is sufficiently distorted. Of course we may derive this from the more basic assumption that the yield function is dependent on $\mathbf{H}$, and that yield is insensitive to superposed rigid motions. This extends to three dimensions the observation, in Section 1.1, to the effect that the uniaxial bar yields when the stress, and hence the elastic stretch, reaches a critical value. We expand the notion of material uniformity by taking the yield function to be the same function at every point p of the materially-uniform body. Thus, yield occurs when the elastic distortion lies on a certain manifold in six-dimensional space. The material is said to respond in the *elastic range* if $\hat{\mathbf{E}}$ is such that $G(\hat{\mathbf{E}}) < 0$, whereas, in rate-independent plasticity theory, the region of strain space where $G(\hat{\mathbf{E}}) > 0$ is deemed to be inaccessible. These statements are modified in the case of rate-dependent, or *viscoplastic*, behavior, to be discussed in Chapters 7 and 9. For metals, the diameter of the elastic range is typically such as to severely limit the norm of the elastic strain, so that as a practical matter we may assume the elastic strain—at least in the rate-independent theory—to be small. This provides post facto justification for the small-elastic-strain assumption invoked in the foregoing.

In these circumstances our constitutive hypotheses imply that the relation between $\hat{\mathbf{E}}$ and $\hat{\mathbf{S}}$ is approximately linear and one-to-one, so that we may equally well characterize yield in terms of the statement $F(\hat{\mathbf{S}}) = 0$, where

$$F(\hat{\mathbf{S}}) = G(\mathcal{L}[\hat{\mathbf{S}}]) \tag{6.89}$$

is the yield function, expressed in terms of the stress. In keeping with the foregoing definition we suppose elastic response to be operative in the elastic range, defined by $F(\hat{\mathbf{S}}) < 0$, and, in the case of rate-independent response, that no state of stress existing in the material can be such that $F(\hat{\mathbf{S}}) > 0$.

We further suppose that $F(\mathbf{0}) < 0$, and hence that the stress-free undistorted state belongs to the elastic range. In this way we partition six-dimensional stress space into the regions defined by positive, negative, and null values of F, with the first of these being inaccessible in any physically possible situation. This appears to disallow behavior of the kind associated with the Bauschinger effect, in which yield can occur upon load reversal before the unloaded state is attained. However, empirical facts support the view that this effect is accompanied by the emergence of dislocations. As we have seen in Section 5.3, these give rise to non-uniform distributions elastic strain, and hence stress, which cannot be directly correlated with the overall global response represented in the test data. From this point of view the Bauschinger effect is thus an artifact of the test being performed, not directly connected with constitutive properties per se. Conventionally, however, the Bauschinger effect is modeled at the constitutive level on the basis of the ad hoc notion of *kinematic hardening*, in which the yield surface translates in stress space according to certain prescribed rules. More recent efforts aimed at modeling the Bauschinger effect take account of the energetic influence of defects, such as the true dislocation density, at the constitutive level. These give rise to space-time partial differential equations for the plastic deformation. Tentative efforts along these lines are described in Chapter 9 in the context of a more expansive discussion of strain hardening.

Returning to the simpler theory, consider a cyclic process in which the deformation and velocity fields start and end at the same values. We assume, as in Chapter 2, that non-negative work must be performed to effect such a process, and thus impose (2.8),

$$\int_{t_1}^{t_2} \mathbf{P} \cdot \dot{\mathbf{F}} dt \geq 0, \tag{6.90}$$

where $t_{1,2}$ respectively are the times when the cycle begins and ends. We henceforth suppress the material point p in the notation—equivalently, its reference position $\mathbf{x}$—as the ensuing discussion pertains to a fixed material point. Suppose the times $t_{1,2}$ are such that the associated stresses satisfy $F < 0$; the cycle begins and ends in the elastic range. If the cycle lies *entirely* within the elastic range, then $\mathbf{K}$ is fixed at the material point in question and we can proceed exactly as in Section 2.1 to conclude, assuming $\mathbf{P}$ to be a function of $\mathbf{F}$, that there exists a family of energy densities $\Psi(\mathbf{F},\mathbf{K})$, say, parametrized by $\mathbf{K}$, such that $\mathbf{P} = \Psi_{\mathbf{F}}(\mathbf{F},\mathbf{K})$, as in $(6.67)_1$.

Problem 6.8. Carry out the details.

Now consider a cycle such that there exists a subinterval of time $[t_a, t_b] \subset [t_1, t_2]$ during which $F = 0$, and that $F < 0$ outside this subinterval. Then we may have plastic flow, i.e., $\dot{\mathbf{K}} \neq \mathbf{0}$ during this subinterval, while $\dot{\mathbf{K}} = \mathbf{0}$ outside it, implying that $\mathbf{K}(t_1) = \mathbf{K}(t_a)$ and $\mathbf{K}(t_2) = \mathbf{K}(t_b)$. Substituting (6.59) and noting that the process is cyclic in the sense that $\mathbf{F}(t_2) = \mathbf{F}(t_1)$, we arrive at the statement

$$\Psi(\mathbf{F}(t_1), \mathbf{K}(t_b)) - \Psi(\mathbf{F}(t_1), \mathbf{K}(t_a)) + \int_{t_a}^{t_b} D dt \geq 0. \tag{6.91}$$

Equivalently,

$$\int_{t_a}^{t_b} [\Psi_{\mathbf{K}}(\mathbf{F}(t_1), \mathbf{K}(t)) \cdot \dot{\mathbf{K}}(t) + D(t)] dt \geq 0. \tag{6.92}$$

To ensure that a cycle beginning in the elastic range $(F(\hat{\mathbf{S}}(t_1)) < 0)$ also ends there, we pass to the limit $t_b - t_a \to 0$. This yields $\mathbf{H}(t_2) = \mathbf{F}(t_2)\mathbf{K}(t_b) = \mathbf{F}(t_1)\mathbf{K}(t_b) \to \mathbf{F}(t_1)\mathbf{K}(t_a) = \mathbf{H}(t_1)$, implying that $\hat{\mathbf{E}}(t_2) \to \hat{\mathbf{E}}(t_1)$ and hence, via the elastic stress–strain relation, that $\hat{\mathbf{S}}(t_2) \to \hat{\mathbf{S}}(t_1)$.

Dividing (6.92) by $t_b - t_a (> 0)$, passing to the limit and invoking the mean-value theorem, we conclude that

$$\Psi_{\mathbf{K}}(\mathbf{F}(t_1), \mathbf{K}(t_a)) \cdot \dot{\mathbf{K}}(t_a) + D(t_a) \geq 0, \tag{6.93}$$

which may be written, using (6.66) and $(6.67)_2$, as

$$[\mathbb{E}(\mathbf{F}(t_a), \mathbf{K}(t_a)) - \mathbb{E}(\mathbf{F}(t_1), \mathbf{K}(t_a))] \cdot \dot{\mathbf{K}}(t_a)\mathbf{K}(t_a)^{-1} \geq 0. \tag{6.94}$$

Alternatively, from (6.69) this inequality may be restated in the form

$$[\mathbb{E}'(\hat{\mathbf{E}}(t_a)) - \mathbb{E}'(\hat{\mathbf{E}}(t_1))] \cdot \mathbf{K}(t_a)^{-1}\dot{\mathbf{K}}(t_a) \geq 0, \tag{6.95}$$

where $\mathbb{E}'(\hat{\mathbf{E}})$ is the function defined by $(6.8)_2$ and (6.73):

$$\mathbb{E}'(\hat{\mathbf{E}}) = U\mathbf{I} - (\mathbf{I} + 2\hat{\mathbf{E}})U_{\hat{\mathbf{E}}}. \tag{6.96}$$

This means that the dissipation is maximized by strains $\hat{\mathbf{E}}$ that lie on the yield surface $G(\hat{\mathbf{E}}) = 0$.

As we have seen, $\mathbb{E}'$ is approximated by $-\hat{\mathbf{S}}$ in the case of small elastic strain. In this case we substitute (6.74), together with $\mathbf{K}^{-1}\dot{\mathbf{K}} = -\dot{\mathbf{G}}\mathbf{G}^{-1}$ (which follows from $\mathbf{GK} = \mathbf{I}$) into (6.95), divide by $\left|\dot{\hat{\mathbf{E}}}\right|$, and pass to the limit $\left|\hat{\mathbf{E}}\right| \to 0$ to derive the leading-order restriction

$$\hat{\mathbf{S}} \cdot Sym\dot{\mathbf{G}}\mathbf{G}^{-1} \geq \hat{\mathbf{S}}^* \cdot Sym\dot{\mathbf{G}}\mathbf{G}^{-1}; \quad F(\hat{\mathbf{S}}^*) \leq 0, \quad F(\hat{\mathbf{S}}) = 0, \tag{6.97}$$

in which the qualifier *Sym* has been inserted to account for the fact that the Piola–Kirchhoff stress is symmetric; the inner product then picks up only the symmetric part of $\dot{\mathbf{G}}\mathbf{G}^{-1}$.

To characterize those $\dot{\mathbf{G}}\mathbf{G}^{-1}$ that satisfy (6.97), consider an arbitrary smooth path $\hat{\mathbf{S}}(u)$, with $u > 0$, such that $\lim_{u \to 0} \hat{\mathbf{S}}(u) = \hat{\mathbf{S}}$ and

$$F(\hat{\mathbf{S}}(u)) \leq 0. \tag{6.98}$$

Then, if the yield function F is differentiable,

$$0 \geq F(\hat{\mathbf{S}}(u)) - F(\hat{\mathbf{S}}) = F_{\hat{\mathbf{S}}}(\hat{\mathbf{S}}) \cdot [\hat{\mathbf{S}}(u) - \hat{\mathbf{S}}] + o(\left|\hat{\mathbf{S}}(u) - \hat{\mathbf{S}}\right|) = uF_{\hat{\mathbf{S}}}(\hat{\mathbf{S}}) \cdot \hat{\mathbf{S}}' + o(u), \tag{6.99}$$

where $\hat{\mathbf{S}}' = \frac{d}{du}\hat{\mathbf{S}}(u)_{|u=0}$. We divide by u and pass to the limit, concluding that

$$F_{\hat{\mathbf{S}}}(\hat{\mathbf{S}}) \cdot \hat{\mathbf{S}}' \leq 0. \tag{6.100}$$

In the same way, it follows from (6.97) that

$$0 \geq Sym\dot{\mathbf{G}}\mathbf{G}^{-1} \cdot [\hat{\mathbf{S}}(u) - \hat{\mathbf{S}}] = u(Sym\dot{\mathbf{G}}\mathbf{G}^{-1}) \cdot \hat{\mathbf{S}}' + o(u), \tag{6.101}$$

and hence that

$$Sym\dot{\mathbf{G}}\mathbf{G}^{-1} \cdot \hat{\mathbf{S}}' \leq 0. \tag{6.102}$$

In the case of equality in (6.100) and (6.102) we have that $Sym\dot{\mathbf{G}}\mathbf{G}^{-1}$ is orthogonal—in the vector space of symmetric tensors—to arbitrary tensors $\hat{\mathbf{S}}'$ that are, in turn, orthogonal to $F_{\hat{\mathbf{S}}}(\hat{\mathbf{S}})$. Thus,

$$Sym\dot{\mathbf{G}}\mathbf{G}^{-1} = \lambda F_{\hat{\mathbf{S}}}, \tag{6.103}$$

for some scalar field $\lambda(\mathbf{x}, t)$, with the derivative evaluated at $\hat{\mathbf{S}}$. Then, $Sym\dot{\mathbf{G}}\mathbf{G}^{-1} \cdot \hat{\mathbf{S}}' = \lambda F_{\hat{\mathbf{S}}} \cdot \hat{\mathbf{S}}'$, and inequalities (6.100) and (6.102) are consistent provided that

$$\lambda \geq 0. \tag{6.104}$$

Taken together, these are the *Kuhn–Tucker* necessary conditions for the optimization problem (6.97). The book by Zangwill may be consulted for further discussion in the context of general optimization theory.

The *flow rule* for the evolution of plastic deformation is thus given by

$$\dot{\mathbf{G}}\mathbf{G}^{-1} = \lambda F_{\hat{\mathbf{S}}} + \mathbf{\Omega}, \tag{6.105}$$

where $\mathbf{\Omega}(\mathbf{x}, t)$ is a skew tensor field, called the *plastic spin.*

With these results in hand (6.97) becomes

$$(\hat{\mathbf{S}} - \hat{\mathbf{S}}^*) \cdot F_{\hat{\mathbf{S}}}(\hat{\mathbf{S}}) \geq 0; \quad F(\hat{\mathbf{S}}^*) \leq 0, \quad F(\hat{\mathbf{S}}) = 0, \tag{6.106}$$

in which $\hat{\mathbf{S}}^*$ is an arbitrary stress in the closure of the elastic range. Thus, the vector $\hat{\mathbf{S}} - \hat{\mathbf{S}}^*$ must either be orthogonal to $F_{\hat{\mathbf{S}}}(\hat{\mathbf{S}})$ or form an acute angle with it. Because $F_{\hat{\mathbf{S}}}(\hat{\mathbf{S}})$ is an exterior normal to the elastic range at the boundary point $\hat{\mathbf{S}}$, this means that its interior, defined by $F < 0$, must lie to one side of the tangent plane T_F at $\hat{\mathbf{S}}$. This tangent plane is unique by virtue of the assumed differentiability of F and the consequent continuity of $F_{\hat{\mathbf{S}}}$. (We will return to the matter of the differentiability of F later.) This, in turn, means that the elastic range is a convex set, i.e., that if $\hat{\mathbf{S}}_1$ and $\hat{\mathbf{S}}_2$ belong to the elastic range, then so do all points on the straight-line path $\hat{\mathbf{S}}(u) = u\hat{\mathbf{S}}_1 + (1 - u)\hat{\mathbf{S}}_2$ with $u \in [0, 1]$ (suggestion: draw a figure). We conclude that the elastic range is always a convex set in a formulation of plasticity theory based on the work inequality. It is important to emphasize, however, that the work inequality itself is an assumption rather than a fundamental law of mechanics, and accordingly that convexity of the elastic range is not a general requirement. It is, however, a basic feature of the classical theory as well as the great majority of its modern generalizations.

Being constitutive in nature, the yield function is naturally subject to material symmetry requirements. Thus, as in (6.27), the yield function $G(\hat{\mathbf{E}})$ must be such that

$$G(\hat{\mathbf{E}}) = G(\mathbf{R}^t\hat{\mathbf{E}}\mathbf{R}), \tag{6.107}$$

for all orthogonal $\mathbf{R} \in g_{\mathcal{K}_i(p)} \cup \{-\mathbf{I}\}$. We have shown that for such $\mathbf{R}$ the stress transforms as indicated in (6.28). Then (6.23) implies that

$$\mathcal{C}[\mathbf{R}^t\hat{\mathbf{E}}\mathbf{R}] = \mathbf{R}^t(\mathcal{C}[\hat{\mathbf{E}}])\mathbf{R}. \tag{6.108}$$

Because $\mathcal{C}$ and $\mathcal{L}$ are mutual inverses, we then have

$$\mathcal{L}[\mathbf{R}^t\hat{\mathbf{S}}\mathbf{R}] = \mathcal{L}[\mathbf{R}^t(\mathcal{C}[\hat{\mathbf{E}}])\mathbf{R}] = \mathcal{L}[\mathcal{C}[\mathbf{R}^t\hat{\mathbf{E}}\mathbf{R}]] = \mathbf{R}^t\hat{\mathbf{E}}\mathbf{R}. \tag{6.109}$$

Combining this with (6.24) we conclude that $G(\mathcal{L}[\hat{\mathbf{S}}]) = G(\mathcal{L}[\mathbf{R}^t\hat{\mathbf{S}}\mathbf{R}])$, and therefore that (6.107) furnishes the restriction

$$F(\hat{\mathbf{S}}) = F(\mathbf{R}^t\hat{\mathbf{S}}\mathbf{R}) \tag{6.110}$$

on the yield function.

Further, invariance of material response under the symmetry transformation $\mathbf{H} \to \mathbf{HR}$ is equivalent to invariance under the transformation $\mathbf{FK} \to \mathbf{FKR}$, which is tantamount to invariance under $\mathbf{K} \to \mathbf{KR}$ at fixed $\mathbf{F}$, or, equivalently, under $\mathbf{G} \to \mathbf{R}^t\mathbf{G}$. Accordingly, $\dot{\mathbf{G}}\mathbf{G}^{-1} \to \mathbf{R}^t(\dot{\mathbf{G}}\,\mathbf{G}^{-1})\mathbf{R}$. This implies that $Sym\dot{\mathbf{G}}\mathbf{G}^{-1} \to \mathbf{R}^t(Sym\dot{\mathbf{G}}\mathbf{G}^{-1})\mathbf{R}$ and hence that the plastic multiplier λ is invariant under material symmetry transformations. We also have $Skw\dot{\mathbf{G}}\mathbf{G}^{-1} \to \mathbf{R}^t(Skw\dot{\mathbf{G}}\mathbf{G}^{-1})\mathbf{R}$ and thus conclude that the plastic spin transforms as

$$\boldsymbol{\Omega} \to \mathbf{R}^t\boldsymbol{\Omega}\mathbf{R}. \tag{6.111}$$

From (6.74), (6.76), and (6.105) the dissipation is given, to leading order in the small elastic strain, by

$$J_K D = \lambda\hat{\mathbf{S}} \cdot F_{\hat{\mathbf{S}}}; \quad F(\hat{\mathbf{S}}) = 0. \tag{6.112}$$

Because $\lambda \geq 0$, the dissipation is positive only if $\lambda > 0$ and hence only if

$$\hat{\mathbf{S}} \cdot F_{\hat{\mathbf{S}}} > 0, \tag{6.113}$$

for all $\hat{\mathbf{S}}$ such that $F(\hat{\mathbf{S}}) = 0$. This should be imposed as an *a priori* restriction on the yield function.

Empirical facts indicate that yield in metals is insensitive to pressure over a very large range of pressures, and certainly for the pressures normally encountered in practice. Thus, yield is insensitive to the value of $tr\mathbf{T}$, where $\mathbf{T}$ is the Cauchy stress. From (6.9) and (6.11) we have that, in the case of small elastic strain,

$$tr\mathbf{T} = tr\hat{\mathbf{S}} + o(|\hat{\mathbf{E}}|). \tag{6.114}$$

Problem 6.9. Prove this.

As the model we are pursuing purports to be valid to leading order in elastic strain, it follows that the yield function should be insensitive to $tr\hat{\mathbf{S}}$. Writing

$$\hat{\mathbf{S}} = Dev\hat{\mathbf{S}} + \tfrac{1}{3}(tr\hat{\mathbf{S}})\mathbf{I}, \tag{6.115}$$

where $Dev\hat{\mathbf{S}}$ is the deviatoric part of $\hat{\mathbf{S}}$, we conclude that the function $\hat{F}(Dev\hat{\mathbf{S}}, tr\hat{\mathbf{S}}) = F(Dev\hat{\mathbf{S}} + \frac{1}{3}(tr\hat{\mathbf{S}})\mathbf{I})$ is insensitive to its second argument, and hence that the yield function depends on $\hat{\mathbf{S}}$ entirely through its deviatoric part. Accordingly, we write

$$F(\hat{\mathbf{S}}) = \tilde{F}(Dev\hat{\mathbf{S}}). \tag{6.116}$$

This assumption imposes a restriction on plastic evolution. To see this we differentiate on the one-parameter family $\hat{\mathbf{S}}(u)$, obtaining

$$F_{\hat{\mathbf{S}}} \cdot \hat{\mathbf{S}}' = \tilde{F}_{Dev\hat{\mathbf{S}}} \cdot (Dev\hat{\mathbf{S}})' = \tilde{F}_{Dev\hat{\mathbf{S}}} \cdot Dev(\hat{\mathbf{S}}') = Dev(\tilde{F}_{Dev\hat{\mathbf{S}}}) \cdot \hat{\mathbf{S}}', \tag{6.117}$$

where $\hat{\mathbf{S}}' = \frac{d}{du}\hat{\mathbf{S}}(u)$, and use has been made of $(Dev\hat{\mathbf{S}})' = Dev(\hat{\mathbf{S}}')$ together with the orthogonality of the deviatoric and spherical tensors, i.e., $Dev\hat{\mathbf{S}} \cdot \mathbf{I} = 0$. It follows that $F_{\hat{\mathbf{S}}} = Dev(\tilde{F}_{Dev\hat{\mathbf{S}}})$ and therefore that $F_{\hat{\mathbf{S}}}$ is deviatoric. Then (6.105) yields the conclusion that plastic flow is isochoric, i.e.,

$$tr(\dot{\mathbf{G}}\mathbf{G}^{-1}) = 0, \tag{6.118}$$

and hence that $(J_G)^{\cdot}$ vanishes. Further, as $Dev(\mathbf{R}^t\hat{\mathbf{S}}\mathbf{R}) = \mathbf{R}^t(Dev\hat{\mathbf{S}})\mathbf{R}$ for any orthogonal $\mathbf{R}$, it follows from (6.110) and (6.116) that

$$\tilde{F}(Dev\hat{\mathbf{S}}) = \tilde{F}(\mathbf{R}^t(Dev\hat{\mathbf{S}})\mathbf{R}). \tag{6.119}$$

We have made essential use of the assumption that the yield function $F(\hat{\mathbf{S}})$ is differentiable. As justification we recall that F, like the strain energy U, is a constitutive function. Having approximated the latter by a quadratic function of $\hat{\mathbf{S}}$ (see (6.25)), logical consistency demands that we also approximate $F(\hat{\mathbf{S}})$ by a quadratic function. Such functions are, of course, continuously differentiable. Because $Dev\hat{\mathbf{S}}$ is a linear function of $\hat{\mathbf{S}}$, it follows that $\tilde{F}(Dev\hat{\mathbf{S}})$ should likewise be approximated by a quadratic function. We take up this issue again in Chapter 7.

The foregoing considerations about yield and flow are quite general and apply to both crystalline and non-crystalline materials.

6.6 Crystallinity versus isotropy

The conventional theory of crystal plasticity rests on a kinematical interpretation of plastic deformation according to which the rate of plastic deformation in a single crystal is presumed to be expressible as a superposition

$$\dot{\mathbf{G}}\mathbf{G}^{-1} = \sum \delta_i \mathbf{s}_i \otimes \mathbf{n}_i \tag{6.120}$$

of simple shear rates, in which $\mathbf{G}$ is the plastic part of the deformation gradient, δ_i are the *slips*, and the $\mathbf{s}_i$ and $\mathbf{n}_i$ are orthonormal vectors specifying the ith slip system. See the books by Havner and Gurtin et al. for a fuller discussion. The sum ranges over the currently active slip systems. Here the δ_i are determined by suitable flow rules, arranged to ensure that the response is dissipative, and the skew part of (6.120), in which the slip-system vectors are specified, furnishes the plastic spin. This decomposition is virtually ubiquitous in the literature on crystal plasticity. However, it has been criticized on the grounds that for finite deformations it cannot be associated with a sequence of simple shears unless these are suitably restricted—see the paper by Rengarajan and Rajagopal. In particular, the order of the sequence generally affects the overall plastic deformation, a fact which is not reflected in (6.120). In the work of Deseri and Owen, conditions are derived under which (6.120) yields an approximation to the deformation associated with a sequence of slips. They find that such deformations are well approximated by (6.120) in face-centered cubic crystals, but the issue remains unresolved for other crystal classes. Further, sequential slip is typical in experiments but simultaneous multi-slip is not usually observed.

This state of affairs provides impetus for alternative phenomenological models based purely on considerations of material symmetry. Indeed, the decoupling of the symmetric and skew parts of $\dot{\mathbf{G}}\mathbf{G}^{-1}$ in the flow rule (6.105) affords considerably more latitude in the fitting of theory to experiment than is possible using (6.120). Accordingly, this is the viewpoint that will be adopted in this book. A drawback, however, relative to a formulation based on (6.120), is that it is then necessary to provide a constitutive specification for the plastic spin.

In crystal-*elasticity* theory the stress arises in response to lattice distortion. The theory is based on the idea that linearly independent, undistorted lattice vectors $\mathbf{l}_i$ $(i \in \{1, 2, 3\})$ are mapped to their images $\mathbf{t}_i$ in κ_t in accordance with the Cauchy–Born hypothesis; that is, the $\mathbf{l}_i$ are convected as material vectors. Chapter 4 of Weiner's book contains an extensive discussion. To accommodate plasticity, this hypothesis is assumed to apply to the *elastic* deformation. Thus, $\mathbf{t}_i = \mathbf{H}\mathbf{l}_i$ where $\mathbf{l}_j$ are the lattice vectors in $\kappa_i(p)$. The lattice set $\{\mathbf{l}_i\}$ associated with $\kappa_i(p)$ is assumed to be an intrinsic property of the crystal. It is therefore regarded as a uniform field (i.e., independent of $\mathbf{x}$) in a single crystal, regarded as a materially uniform body.

The $\mathbf{t}_i$ are observable in principle. In practice they are computed from their reciprocals $\mathbf{t}^i$, the normals to lattice planes, which are measured in X-ray diffraction experiments. The decomposition (5.3) yields $\mathbf{t}_i = \mathbf{F}\mathbf{r}_i$, where $\mathbf{r}_i = \mathbf{K}\mathbf{l}_i$ are the lattice vectors in

κ. The plastic deformation is then given by $\mathbf{K} = \mathbf{r}_i \otimes \mathbf{l}^i$, where the $\mathbf{l}^i$ are the reciprocals of the $\mathbf{l}_j$. The elastic deformation is $\mathbf{H} = \mathbf{t}_i \otimes \mathbf{l}^i$, and the deformation gradient is $\mathbf{F} = \mathbf{t}_i \otimes \mathbf{r}^i$.

The material derivatives of the referential lattice vectors are $\dot{\mathbf{r}}_i = \dot{\mathbf{K}}\mathbf{l}_i + \mathbf{K}\dot{\mathbf{l}}_i$. These imply that if $\dot{\mathbf{l}}_i \neq \mathbf{0}$, then the lattice vectors are non-material ($\dot{\mathbf{r}}_i \neq \mathbf{0}$) in the absence of plastic flow ($\dot{\mathbf{K}} = \mathbf{0}$). However, we adopt the prevalent view that plastic flow is solely responsible for the non-materiality of the lattice, i.e., that plastic flow alone accounts for the evolution of material vectors relative to the lattice. We thus impose $\dot{\mathbf{l}}_i = \mathbf{0}$ and regard the set $\{\mathbf{l}_i\}$ of lattice vectors as assigned data. This, in turn, yields the materiality of the set $\{\mathbf{r}_i\}$ in the absence of plastic flow, in accordance with the conventional statement of the Cauchy–Born hypothesis for purely elastic deformations. Note the contrast with the representations (4.9), (5.24), and (5.47), for the deformation gradient and its elastic and plastic factors, in which the $\mathbf{e}_i$ are material vectors, i.e., $\dot{\mathbf{e}}_i = \mathbf{0}$.

The present interpretation comports with (6.120) in which the slip-system vectors are considered to be fixed; these, in turn, may be determined once the set $\{\mathbf{l}_i\}$ is specified. Further, for the purpose of integrating the flow rule it is necessary to assign an initial value of plastic deformation at time t_0, say, and this is another reason why it is necessary to specify the lattice $\{\mathbf{l}_i\}$. If we choose $\kappa = \kappa_{t_0}$, for example, then because the referential lattice coincides with the actual lattice at time t_0 and is therefore measurable in principle, the specification of $\{\mathbf{l}_i\}$ provides the initial value of $\mathbf{K}$, and hence that of $\mathbf{G}$.

The fact that the lattice $\{\mathbf{l}_i\}$ is fixed, independent of $\mathbf{x}$ and t, implies that the elements $\mathbf{R}$ of the (discrete) material symmetry group $g_{\kappa_i(p)}$ are also fixed, i.e., invariant in time and the same at all material points of the crystal. This has important consequences for the further development of the theory.

Naturally, none of the foregoing considerations concerning lattice vectors apply to isotropic solids, for which $g_{\kappa_i(p)} \cup \{-\mathbf{I}\} = Orth$. Thus, for these there is no requirement that $\mathbf{R} \in g_{\kappa_i(p)}$ be uniformly distributed, or even that it be independent of time. We exploit this observation, in Chapter 7, to simplify the theory for isotropic solids accordingly.

Problem 6.10. Consider a crystalline solid having cubic symmetry relative to $\kappa_i(p)$. Let the edges of the cube be aligned with an orthonormal basis $\{\mathbf{l}_i\}$. It is known—see, for example, the book by Green and Adkins—that the most general homogeneous quadratic strain-energy function for such a solid is a linear combination of

$$(E_{11} + E_{22} + E_{33})^2, \quad E_{11}E_{22} + E_{11}E_{33} + E_{22}E_{33} \quad \text{and} \quad E_{12}^2 + E_{13}^2 + E_{23}^2,$$

where $E_{ij} = \hat{\mathbf{E}} \cdot Sym(\mathbf{l}_i \otimes \mathbf{l}_j)$ are the Cartesian components of the strain $\hat{\mathbf{E}}$.

(a) Show that the general quadratic strain-energy function for a uniform cubic solid is thus expressible in the form

$$U(\hat{\mathbf{E}}) = \tfrac{1}{2}[C_1(E_{11} + E_{22} + E_{33})^2 + C_2(\bar{E}_{11}^2 + \bar{E}_{22}^2 + \bar{E}_{33}^2) + C_3(E_{12}^2 + E_{13}^2 + E_{23}^2)],$$

where C_i are constants (the elastic moduli) and $\bar{E}_{ij}$ are the components of $Dev\hat{\mathbf{E}}$.

(b) Show that $U(\hat{\mathbf{E}})$ is positive definite if and only if all $C_i > 0$.

(c) Derive expressions for the Cartesian components S_{ij} of the Piola-Kirchhoff stress $\hat{\mathbf{S}} = U_{\hat{\mathbf{E}}}$ relative to the basis $\{\mathbf{l}_i \otimes \mathbf{l}_j\}$.

6.7 Discontinuous fields

It is of interest to extend the notion of dissipation to accommodate discontinuities in the deformation gradient, to account for the dissipation attending the evolution of phase boundaries. To this end we first extend the balance laws to allow for such discontinuities. A basic tool in this endeavor is the extension of Reynolds' transport theorem to non-material subregions of the body. Thus, consider a region $R_t \subset \kappa_t$ and let $v_n(\mathbf{y}, t)$ be the normal speed of its boundary ∂R_t in the direction of its exterior unit normal. Then for any smooth scalar field $\varphi(\mathbf{y}, t)$,

$$\frac{d}{dt}\int_{R_t} \varphi dv = \int_{R_t} \varphi_t dv + \int_{\partial R_t} \varphi v_n da, \tag{6.121}$$

where $\varphi_t = \partial\varphi(\mathbf{y}, t)/\partial t$. See Chapter 2 of Liu's book for an elementary proof. If R_t should happen to be a *material* region π_t, say, i.e., a region that evolves in such a way as to always contain the same set of material points, then $v_n = \mathbf{v} \cdot \mathbf{n}$, where $\mathbf{v}(\mathbf{y}, t) = \dot{\mathbf{y}}$ is the material velocity and $\mathbf{n}(\mathbf{y}, t)$ is the exterior unit normal to $\partial\pi_t$. We thus recover the conventional statement of the transport theorem:

$$\frac{d}{dt}\int_{\pi_t} \varphi dv = \int_{\pi_t} \varphi_t dv + \int_{\partial\pi_t} \varphi \mathbf{v} \cdot \mathbf{n} da. \tag{6.122}$$

Similarly, if $R(t) \subset \kappa$ is an evolving, and hence non-material, subvolume of the chosen reference configuration, with outward normal speed $V_\nu(\mathbf{x}, t)$, then

$$\frac{d}{dt}\int_{R} \Phi dV = \int_{R} \dot{\Phi} dV + \int_{\partial R} \Phi V_\nu dA, \tag{6.123}$$

for smooth fields $\Phi(\mathbf{x}, t)$; and, if $R = \pi$, a material region, then, because $\dot{\mathbf{x}}$ vanishes by definition,

$$\frac{d}{dt}\int_{\pi} \Phi dV = \int_{\pi} \dot{\Phi} dV. \tag{6.124}$$

Consider now a field Φ that suffers a discontinuity across a non-material surface $S \subset \kappa$. Consider two disjoint material subvolumes π_1, π_2 in κ, and suppose S traverses π_2, dividing it into two non-material regions $R_\pm$, so that $\pi_2 = R_+ \cup R_-$. Let $(\partial\pi_2)_\pm = \partial R_\pm \cap \partial\pi_2$. Then, $\partial\pi_2 = (\partial\pi_2)_+ \cup (\partial\pi_2)_-$. It may be helpful to draw a figure. Finally, let $s = \pi_2 \cap S$; this is the part of S contained in π_2. If Φ is smooth in the separate regions

$R_\pm$, and if V_ν is the normal speed of S in the direction of the unit normal $\boldsymbol{\nu}_S$ to s directed from R_- to R_+, then it follows from (6.123) that

$$\frac{d}{dt}\int_{R_\pm} \Phi dV = \int_{R_\pm} \dot{\Phi} dV \mp \int_s \Phi_\pm V_\nu dA, \tag{6.125}$$

where $\Phi_\pm$, respectively, are the limits of Φ as s is approached from $R_\pm$. Adding these two equations then yields

$$\frac{d}{dt}\int_\pi \Phi dV = \int_\pi \dot{\Phi} dV - \int_s [\Phi] V_\nu dA, \quad \text{where} \quad [\Phi] = \Phi_+ - \Phi_- \tag{6.126}$$

is the jump of Φ across s and $\pi = \pi_2$. On the other hand, (6.124) holds as it stands if $\pi = \pi_1$.

In the same way, if the Piola stress $\mathbf{P}(\mathbf{x},t)$ is smooth in $R_\pm$, then the divergence theorem, applied to each of these subregions, yields, upon adding the two statements,

$$\int_{\partial\pi} \mathbf{P}\boldsymbol{\nu} dA = \int_\pi Div\mathbf{P} dV + \int_s [\mathbf{P}]\boldsymbol{\nu}_S dA. \tag{6.127}$$

This holds if $\pi = \pi_2$, whereas the same statement, but with the integral over s omitted, holds if $\pi = \pi_1$. The conventional statement of the divergence theorem is similarly generalized as

$$\int_{\partial\pi} \mathbf{a}\cdot\boldsymbol{\nu} dA = \int_\pi Div\mathbf{a} dV + \int_s [\mathbf{a}]\cdot\boldsymbol{\nu}_S dA, \tag{6.128}$$

where $\mathbf{a}(\mathbf{x}, t)$ is any piecewise smooth vector field; and, for any such scalar field $\Phi(\mathbf{x}, t)$, as

$$\int_{\partial\pi} \Phi\boldsymbol{\nu} dA = \int_\pi \nabla\Phi dV + \int_s [\Phi]\boldsymbol{\nu}_S dA. \tag{6.129}$$

Applying the latter to the components of $\mathbf{a}$ on a fixed basis and adding the results yields

$$\int_{\partial\pi} \mathbf{a}\otimes\boldsymbol{\nu} dA = \int_\pi \nabla\mathbf{a} dV + \int_s [\mathbf{a}]\otimes\boldsymbol{\nu}_S dA. \tag{6.130}$$

In particular, then, if the deformation $\boldsymbol{\chi}(\mathbf{x}, t)$ is continuous,

$$\int_\pi \mathbf{F} dV = \int_{\partial\pi} \boldsymbol{\chi}\otimes\boldsymbol{\nu} dA, \tag{6.131}$$

where, of course, $\mathbf{F} = \nabla\boldsymbol{\chi}$ is the deformation gradient.

As an example consider the partwise linear momentum balance (1.24), expressed in referential form, i.e.,

$$\tfrac{d}{dt}\int_{\pi}\rho_{\kappa}\mathbf{v}dV=\int_{\pi}\rho_{\kappa}\mathbf{b}dV+\int_{\partial\pi}\mathbf{P}\boldsymbol{\nu}dA. \tag{6.132}$$

This makes sense even for discontinuous fields and so should be regarded as the fundamental form of the balance statement. Applying (6.126) to the left-hand side, we reduce this to

$$\int_{\partial\pi}\mathbf{P}\boldsymbol{\nu}dA+\int_{\pi}\rho_{\kappa}(\mathbf{b}-\dot{\mathbf{v}})dV=-\int_{s}[\rho_{\kappa}\mathbf{v}]V_{\nu}dA. \tag{6.133}$$

For $\pi=\pi_1$ the right-hand side of this expression vanishes and we can proceed as in Chapter 1 to conclude that this is equivalent to the local equation of motion (1.40). If $\pi=\pi_2$ then the latter applies in the separate regions $R_{\pm}$.

Consider a sequence of regions π_2 that collapse onto s. Assuming $|\rho_{\kappa}(\mathbf{b}-\dot{\mathbf{v}})|$ to be bounded in $R_{\pm}$, (6.133) reduces, in the limit, to

$$\int_{s}\{\mathbf{P}_{+}\boldsymbol{\nu}_{S}+\mathbf{P}_{-}(-\boldsymbol{\nu}_{S})\}dA+\int_{s}[\rho_{\kappa}\mathbf{v}]V_{\nu}dA=\mathbf{0}. \tag{6.134}$$

As π_2 is arbitrary and hence so too its intersection s with S, we may localize and conclude that

$$[\mathbf{P}]\boldsymbol{\nu}_{S}+[\rho_{\kappa}\mathbf{v}]V_{\nu}=\mathbf{0}\quad\text{on}\quad S. \tag{6.135}$$

As a second example consider the global mass conservation law

$$\tfrac{d}{dt}\int_{\pi}\rho_{\kappa}dV=0. \tag{6.136}$$

If $\pi=\pi_1$, then, as we have seen in Chapter 1, we may localize and conclude that $\dot{\rho}_{\kappa}=0$ pointwise, which of course also holds in $R_{\pm}$ in the case when $\pi=\pi_2$. Applying (6.126) and (6.136) directly to π_2, we have

$$\int_{\pi}\dot{\rho}_{\kappa}dV-\int_{s}[\rho_{\kappa}]V_{\nu}dA=0. \tag{6.137}$$

Because $\dot{\rho}_{\kappa}$ vanishes in $\pi\setminus s$ this reduces to

$$\int_{s}[\rho_{\kappa}]V_{\nu}dA=0, \tag{6.138}$$

which we again localize, concluding that

$$[\rho_{\kappa}]V_{\nu}=0\quad\text{on}\quad S. \tag{6.139}$$

Accordingly, if S is not a material surface, i.e., if $V_\nu \neq 0$, then $[\rho_\kappa] = 0$ and (6.135) becomes

$$[\mathbf{P}]\boldsymbol{\nu}_S + \rho_\kappa V_\nu [\mathbf{v}] = \mathbf{0} \quad \text{on} \quad S. \tag{6.140}$$

This agrees with (6.135) when S *is* a material surface, and is thus valid in general.

Next, we apply (6.126) with $\boldsymbol{\Phi}$ replaced by the deformation gradient:

$$\tfrac{d}{dt}\int_\pi \mathbf{F}dV = \int_\pi \dot{\mathbf{F}}dV - \int_s [\mathbf{F}]V_\nu dA. \tag{6.141}$$

To reduce this we differentiate (6.131) and substitute into the left-hand side to obtain

$$\int_{\partial\pi} \mathbf{v} \otimes \boldsymbol{\nu} dA = \int_\pi \dot{\mathbf{F}}dV - \int_s [\mathbf{F}]V_\nu dA, \tag{6.142}$$

where $\mathbf{v} = \dot{\chi}$ is the material velocity. Collapsing π onto s and assuming $|\dot{\mathbf{F}}|$ to be bounded then yields

$$\int_s \{\mathbf{v}_+ \otimes \boldsymbol{\nu}_S + \mathbf{v}_- \otimes (-\boldsymbol{\nu}_S)\}dA + \int_s [\mathbf{F}]V_\nu dA, \tag{6.143}$$

which we localize as usual to conclude that

$$[\mathbf{v}] \otimes \boldsymbol{\nu}_S = -[\mathbf{F}]V_\nu \quad \text{on} \quad S. \tag{6.144}$$

Combining this with (6.53), i.e., $[\mathbf{F}] = \mathbf{f} \otimes \boldsymbol{\nu}_S$, then yields

$$[\mathbf{v}] = -V_\nu \mathbf{f} \quad \text{on} \quad S. \tag{6.145}$$

We are finally in a position to examine the dissipation. Recall that this is subject to inequality (6.57), where $\mathcal{D}$ is defined in (6.56). This statement makes sense for fields that are smooth or not, provided the associated integrals exist. For our present purposes it is convenient to write the total energy $\mathcal{U} + \mathcal{K}$ in the form

$$\mathcal{U}(\pi, t) + \mathcal{K}(\pi, t) = \int_\pi \Phi dV, \quad \text{where} \quad \Phi = \Psi + \tfrac{1}{2}\rho_\kappa |\mathbf{v}|^2 \tag{6.146}$$

is the energy density. Then,

$$\mathcal{D}(\pi, t) = \mathcal{P}(\pi, t) - \tfrac{d}{dt}\int_\pi \Phi dV, \tag{6.147}$$

where $\mathcal{P}(\pi, t)$, given by (1.47), is the power supply. Using (6.128), we reduce the latter to

$$\mathcal{P}(\pi, t) - \int_{\pi} \rho_{\kappa}\mathbf{b} \cdot \mathbf{v}dV = \int_{\partial\pi} \mathbf{P}^t\mathbf{v} \cdot \boldsymbol{\nu}dA = \int_{\pi} Div(\mathbf{P}^t\mathbf{v})dV + \int_{s}[\mathbf{P}^t\mathbf{v}] \cdot \boldsymbol{\nu}_S dA, \quad (6.148)$$

in which the integrals over π are to be regarded as integrals over $\pi \setminus s$, the two sets having the same volume measure. In the latter (1.40) is valid, and may be scalar-multiplied by $\mathbf{v}$ to obtain

$$Div(\mathbf{P}^t\mathbf{v}) + \rho_{\kappa}\mathbf{b} \cdot \mathbf{v} = \dot{\Phi} + D, \quad (6.149)$$

where D is the volumetric dissipation density given by (6.59) and (6.66). Thus,

$$\mathcal{P}(\pi, t) = \int_{\pi} (\dot{\Phi} + D)dV + \int_{s}[\mathbf{P}^t\mathbf{v}] \cdot \boldsymbol{\nu}_S dA, \quad (6.150)$$

which combines with (6.126) and (6.147) to give

$$\mathcal{D}(\pi, t) = \int_{\pi} DdV + \int_{s} D_S dA, \quad (6.151)$$

where

$$D_S = [\mathbf{P}^t\mathbf{v}] \cdot \boldsymbol{\nu}_S + V_{\nu}[\Phi] \quad (6.152)$$

is the surface density of dissipation.

Problem 6.11. Prove these statements.

To further reduce (6.152) we introduce the arithmetic mean $\langle\cdot\rangle$ of the limiting values of a function on either side of a surface of discontinuity. Then for any functions f and g,

$$[fg] = [f]\langle g\rangle + \langle f\rangle[g]. \quad (6.153)$$

Accordingly,

$$\begin{aligned} [\mathbf{P}^t\mathbf{v}] \cdot \boldsymbol{\nu}_S &= ([\mathbf{P}^t]\langle\mathbf{v}\rangle + \langle\mathbf{P}^t\rangle[\mathbf{v}]) \cdot \boldsymbol{\nu}_S = [\mathbf{P}]\boldsymbol{\nu}_S \cdot \langle\mathbf{v}\rangle + [\mathbf{v}] \cdot \langle\mathbf{P}\rangle\boldsymbol{\nu}_S \\ &= [\mathbf{P}]\boldsymbol{\nu}_S \cdot \langle\mathbf{v}\rangle + \langle\mathbf{P}\rangle \cdot [\mathbf{v}] \otimes \boldsymbol{\nu}_S, \end{aligned} \quad (6.154)$$

and (6.140), (6.145), together with $[\mathbf{v}] \cdot \langle\mathbf{v}\rangle = \frac{1}{2}[|\mathbf{v}|^2]$, result in

$$[\mathbf{P}^t\mathbf{v}] \cdot \boldsymbol{\nu}_S = -\tfrac{1}{2}V_{\nu}[\rho_{\kappa}|\mathbf{v}|^2] - V_{\nu}\langle\mathbf{P}\rangle \cdot [\mathbf{F}]. \quad (6.155)$$

Using $[\mathbf{F}] = \mathbf{f} \otimes \boldsymbol{\nu}_S$ with $\mathbf{f} = [\mathbf{F}]\boldsymbol{\nu}_S$, i.e., $[\mathbf{F}] = [\mathbf{F}]\boldsymbol{\nu}_S \otimes \boldsymbol{\nu}_S$, together with (6.140), (6.145), and the easily derived rule $\mathbf{A} \cdot \mathbf{a} \otimes \mathbf{b} = \mathbf{a} \cdot \mathbf{A}\mathbf{b}$, we have

$$\langle \mathbf{P} \rangle \cdot [\mathbf{F}] = [\mathbf{F}]\boldsymbol{\nu}_S \cdot \langle \mathbf{P} \rangle \boldsymbol{\nu}_S = \boldsymbol{\nu}_S \cdot [\mathbf{F}^t]\langle \mathbf{P} \rangle \boldsymbol{\nu}_S = \boldsymbol{\nu}_S \cdot ([\mathbf{F}^t\mathbf{P}] - \langle \mathbf{F}^t \rangle [\mathbf{P}])\boldsymbol{\nu}_S, \tag{6.156}$$

in which, again by (6.140) and (6.145),

$$[\mathbf{P}]\boldsymbol{\nu}_S = \rho_\kappa V_\nu^2 [\mathbf{F}]\boldsymbol{\nu}_S. \tag{6.157}$$

Now, (6.153) gives $\langle \mathbf{F}^t \rangle [\mathbf{F}] = [\mathbf{F}^t\mathbf{F}] - [\mathbf{F}^t]\langle \mathbf{F} \rangle$, so that $\boldsymbol{\nu}_S \cdot \langle \mathbf{F}^t \rangle [\mathbf{F}]\boldsymbol{\nu}_S = \boldsymbol{\nu}_S \cdot [\mathbf{F}^t\mathbf{F}]\boldsymbol{\nu}_S - \boldsymbol{\nu}_S \cdot [\mathbf{F}]^t \langle \mathbf{F} \rangle \boldsymbol{\nu}_S$, where $\boldsymbol{\nu}_S \cdot [\mathbf{F}]^t \langle \mathbf{F} \rangle \boldsymbol{\nu}_S = [\mathbf{F}]\boldsymbol{\nu}_S \cdot \langle \mathbf{F} \rangle \boldsymbol{\nu}_S = \boldsymbol{\nu}_S \cdot \langle \mathbf{F}^t \rangle [\mathbf{F}]\boldsymbol{\nu}_S$. This gives $\boldsymbol{\nu}_S \cdot \langle \mathbf{F}^t \rangle [\mathbf{F}]\boldsymbol{\nu}_S = \frac{1}{2}\boldsymbol{\nu}_S \cdot [\mathbf{F}^t\mathbf{F}]\boldsymbol{\nu}_S$, and (6.152) is finally reduced to

$$D_S = V_\nu \boldsymbol{\nu}_S \cdot ([\mathbb{E}] + \tfrac{1}{2}\rho_\kappa V_\nu^2 [\mathbf{F}^t\mathbf{F}])\boldsymbol{\nu}_S, \tag{6.158}$$

where $\mathbb{E}$ is the Eshelby tensor, defined by (6.65). Note that D_S vanishes if S is a material surface ($V_\nu = 0$).

The global dissipation inequality thus requires that

$$\int_\pi D\,dV + \int_s D_S\,dA \geq 0 \quad \text{for all} \quad \pi \subset \kappa. \tag{6.159}$$

If $\pi = \pi_1$, then $s = \emptyset$ and localization yields the pointwise inequality $D \geq 0$, as before. Naturally this also holds in $R_\pm$ if $\pi = \pi_2$. Collapsing π onto s results in

$$\int_s D_S\,dA \geq 0 \quad \text{for all} \quad s \subset S, \tag{6.160}$$

assuming, of course, that D is bounded. It follows, by localizing as usual, that

$$D_S \geq 0 \quad \text{on} \quad S, \tag{6.161}$$

and (6.158) once again highlights the central role played by the Eshelby tensor in controlling the dissipation. The latter equation indicates that the function $F(V_\nu)$, obtained by fixing all variables other than V_ν therein, then satisfies $F(V_\nu) \geq 0$ and $F(0) = 0$. Thus, F is minimized at $V_\nu = 0$, yielding the *phase equilibrium* condition at a stationary phase boundary:

$$\boldsymbol{\nu}_S \cdot [\mathbb{E}]\boldsymbol{\nu}_S = 0 \quad \text{on} \quad S, \quad \text{when} \quad V_\nu = 0. \tag{6.162}$$

See the paper by Abeyaratne and Knowles and the book by Liu for derivations of (6.158) and (6.162) in a more general thermodynamical setting.

Problem 6.12. Fill in the steps leading to these two equations.

With these results in hand we may conclude, with reference to Section 2.2 and in the case of conservative problems, that plastic evolution furnishes the dissipation needed to ensure that asymptotically stable equilibria minimize the total potential energy. Thus, suppose there exists a trajectory $\{\chi(\mathbf{x},t),\mathbf{K}(\mathbf{x},t)\}$ with

$$\{\chi(\mathbf{x},t_0),\mathbf{K}(\mathbf{x},t_0)\} = \{\chi_0(\mathbf{x}),\mathbf{K}_0(\mathbf{x})\} \tag{6.163}$$

and

$$\lim_{t\to\infty}\{\chi(\mathbf{x},t),\mathbf{K}(\mathbf{x},t)\} = \{\chi_\infty(\mathbf{x}),\mathbf{K}_\infty(\mathbf{x})\}, \tag{6.164}$$

with

$$\dot{\chi}(\mathbf{x},t)_{|t_0} = \mathbf{0} \quad \text{and} \quad \lim_{t\to\infty}\dot{\chi}(\mathbf{x},t) = \mathbf{0}. \tag{6.165}$$

Then, because $\mathcal{D}(\kappa,t) \geq 0$ in the presence of plastic evolution, (2.42) obtains and yields (2.45) in the form

$$\mathcal{E}[\chi_\infty,\mathbf{K}_\infty] \leq \mathcal{E}[\chi_0,\mathbf{K}_0], \tag{6.166}$$

where

$$\mathcal{E}[\chi,\mathbf{K}] = \int_\kappa \Psi(\nabla\chi,\mathbf{K})dV - \mathcal{L} \tag{6.167}$$

is the total potential energy in which $\mathcal{L}$ is the relevant load potential.

References

Abeyaratne, R., and Knowles, J. K. (1990). On the driving traction acting on a surface of strain discontinuity in a continuum. *J. Mech. Phys. Solids* 38, 345–60.

Batchelor, G. K. (Ed) (1958). *The Scientific Papers of Sir Geoffrey Ingram Taylor*, Vol. 1: *Mechanics of Solids*. Cambridge University Press, Cambridge, UK.

Deseri, L., and Owen, D. R. (2002). Invertible structured deformations and the geometry of multiple slip in single crystals. *Int. J. Plasticity* 18, 833–49.

Epstein, M., and Maugin, G. A. (1990). The energy-momentum tensor and material uniformity in finite elasticity. *Acta Mechanica* 83, 127–33.

Green, A. E., and Adkins, J. E. (1970). *Large Elastic Deformations*. Oxford University Press, Oxford.

Gurtin, M. E., Fried, E., and Anand, L. (2010). *The Mechanics and Thermodynamics of Continua*. Cambridge University Press, Cambridge, UK.

Havner, K. S. (1992). *Finite Plastic Deformation of Crystalline Solids.* Cambridge University Press, Cambridge, UK.
Lardner, R. W. (1974). *Mathematical Theory of Dislocations and Fracture.* University of Toronto Press, Toronto.
Liu, I-Shih. (2002). *Continuum Mechanics.* Springer, Berlin.
Nabarro, F. R. N. (1987). *Theory of Dislocations.* Dover, New York.
Noll, W. (1967). Materially uniform simple bodies with inhomogeneities. *Arch. Ration. Mech. Anal.* 27, 1–32.
Rengarajan, G., and Rajagopal, K. R. (2001). On the form for the plastic velocity gradient L_p in crystal plasticity. *Math. Mech. Solids* 6, 471–80.
Truesdell, C., and Toupin, R. A. (1960). The classical field theories. In: Flügge, S. (Ed.), *Handbuch der Physik*, Vol. III/1, pp. 226–793. Springer, Berlin.
Weiner, J. H. (2002). *Statistical Mechanics of Elasticity.* Dover, New York.
Zangwill, W. I. (1969). *Nonlinear Programming.* Prentice-Hall, Englewood Cliffs, NJ.

7

Isotropy

The theory for isotropic solids is by far the most highly developed and most widely applied branch of plasticity. Its dominant position is justified by the fact that most engineering metals are polycrystalline, with the undistorted crystal grains oriented more or less randomly on the mesoscale, so that, at this scale, the solid responds in the manner of an isotropic material. Thus, the theory for isotropic solids covers the large majority of engineering applications, and for this reason we devote considerable space to it here.

We can imagine that as the solid is deformed, the individual grains reorient themselves accordingly, and hence that a texture is developed in the material in response to plastic flow. We might thus conceive of a more general framework than that which we have considered thus far, one in which the basic kinematic descriptors are independent deformation and rotation fields, both assigned to the material points of the continuum, with the latter representing the mesoscopic-level grain orientation. This would interact with the conventional deformation in accordance with appropriate balance laws and constitutive equations. The resulting model would fall under the umbrella of the theory of Cosserat continua, a well-established framework conceived precisely to account for rotational degrees of freedom not taken into account by the conventional Cauchy continuum. Section 98 of the treatise by Truesdell and Noll provides a good survey of the theory of Cosserat continua. Though the extension of such a model to plasticity would encounter substantial obstacles, some progress along these lines has recently been made. We refer the interested reader to the paper by Neff and the book by Epstein and Elżanowski for further discussion. We shall not delve into this model here, however, as it has yet to reach a level of maturity that would justify its inclusion in a course. Instead, we will focus on the standard theory, which remains the primary model to this day, despite its limitations.

7.1 The flow rule

To begin, we recall that the yield function satisfies the material symmetry condition (6.110), in which $\mathbf{R}$ is an orthogonal tensor belonging to the symmetry group.

A Course on Plasticity Theory. David J. Steigmann, Oxford University Press. © David J. Steigmann (2022).
DOI: 10.1093/oso/9780192883155.003.0007

This restriction is entirely similar to the condition (6.27) satisfied by the strain-energy function. We may thus proceed, as in the passage from the latter to (6.34), to conclude that

$$F_{\hat{S}}(\mathbf{R}^t\hat{\mathbf{S}}\mathbf{R}) = \mathbf{R}^t F_{\hat{S}}(\hat{\mathbf{S}})\mathbf{R}. \tag{7.1}$$

This, of course, is valid in general, not only in the case of isotropy. However, in the case of isotropy it may be used to effect a major simplification of the flow rule.

To this end, recall that invariance of the strain energy or yield function under material symmetry transformations is tantamount to their invariance under the replacement of $\mathbf{H}$ by $\bar{\mathbf{H}} = \mathbf{HR}$, where, in the case of isotropy, $\mathbf{R}$ is any rotation whatsoever. From (5.3), this is equivalent to invariance under replacement of $\mathbf{K}$ by $\bar{\mathbf{K}} = \mathbf{KR}$—equivalently, replacement of $\mathbf{G}$ by $\bar{\mathbf{G}} = \mathbf{R}^t\mathbf{G}$—with $\mathbf{F}$ remaining fixed. To see how this replacement affects plastic flow, we note that

$$(\bar{\mathbf{G}})^{\cdot}\bar{\mathbf{G}}^{-1} = \mathbf{R}^t\dot{\mathbf{G}}\mathbf{G}^{-1}\mathbf{R} + \dot{\mathbf{R}}^t\mathbf{R}, \tag{7.2}$$

where, recalling the discussion in Section 6.6, we have allowed for the possibility that $\mathbf{R}$ may be time-dependent. In particular, the argument used there to justify the conclusion that $\mathbf{R}$ is fixed in the case of crystalline response is not applicable in the case of isotropy. Substituting (6.105), we conclude that

$$(\bar{\mathbf{G}})^{\cdot}\bar{\mathbf{G}}^{-1} = \lambda\mathbf{R}^t F_{\hat{S}}(\hat{\mathbf{S}})\mathbf{R} + \mathbf{R}^t(\boldsymbol{\Omega} + \mathbf{R}\dot{\mathbf{R}}^t)\mathbf{R}, \tag{7.3}$$

where $\boldsymbol{\Omega}$ is the plastic spin.

Because $\boldsymbol{\Omega}$ is skew we can always find a rotation $\mathbf{R}(t)$ to nullify the parenthetical term. As usual, we suppress the passive argument $\mathbf{x}$ in the notation as we are concerned with a fixed material point. Thus, suppose $\mathbf{B}(t)$ satisfies the initial-value problem

$$\dot{\mathbf{B}} = \boldsymbol{\Omega}\mathbf{B} \quad \text{with} \quad \mathbf{B}(0) = \mathbf{B}_0, \tag{7.4}$$

where $\mathbf{B}_0$ is a rotation. Let $\mathbf{Z} = \mathbf{B}\mathbf{B}^t$; then,

$$\dot{\mathbf{Z}} = \boldsymbol{\Omega}\mathbf{Z} - \mathbf{Z}\boldsymbol{\Omega}, \quad \text{with} \quad \mathbf{Z}(0) = \mathbf{I}. \tag{7.5}$$

The unique solution is $\mathbf{Z}(t) = \mathbf{I}$, implying that $\mathbf{B}$ is orthogonal. Further,

$$\dot{\mathcal{J}}_B = \mathbf{B}^* \cdot \dot{\mathbf{B}} = \mathcal{J}_B tr(\dot{\mathbf{B}}\mathbf{B}^{-1}) = \mathcal{J}_B tr\boldsymbol{\Omega}, \tag{7.6}$$

and this vanishes because $\boldsymbol{\Omega}$ is skew. Then, $\mathcal{J}_{B(t)} = \mathcal{J}_{B(0)} = 1$ and $\mathbf{B}$ is a rotation. Because the rotation $\mathbf{R}$ in (7.3) is arbitrary, we are free to pick $\mathbf{R} = \mathbf{B}$ and thus obtain

$$(\bar{\mathbf{G}})^{\cdot}\bar{\mathbf{G}}^{-1} = \bar{\lambda}F_{\bar{S}}(\bar{\mathbf{S}}), \quad \text{with} \quad \bar{\lambda} = \lambda \quad \text{and} \quad \bar{\mathbf{S}} = \mathbf{B}^t\hat{\mathbf{S}}\mathbf{B}. \tag{7.7}$$

Problem 7.1. For isotropic materials, show that the Cauchy stress $\mathbf{T}$ is invariant under the replacement $\mathbf{H} \to \bar{\mathbf{H}} = \mathbf{HB}$. Because $\mathbf{F}$ is invariant, this means that the Piola stress $\mathbf{P}$ is likewise invariant.

Accordingly, use of the flow rule (7.7), with initial condition $\bar{\mathbf{G}}(0) = \mathbf{B}_0^t\mathbf{G}(0)$, yields the same deformation and stress fields as would be obtained by using the original flow rule (6.105). In effect, then, for isotropic materials we can exploit the degree of freedom afforded by the material symmetry group to suppress plastic spin in the flow rule altogether. This major simplification is not possible in crystalline materials.

Problem 7.2. Why not?

Because $\bar{\mathbf{G}}$ and $\mathbf{G}$ are mechanically indistinguishable in the case of isotropy, we may, without loss of generality, identify the former with the plastic deformation and regard $\bar{\mathbf{H}}$ as the elastic deformation. Then, to ease the notation, we simply drop the overbars and write (7.7) as

$$\dot{\mathbf{G}}\mathbf{G}^{-1} = \lambda F_{\hat{\mathbf{S}}}. \tag{7.8}$$

7.2 Von Mises' yield function

Recall the material symmetry condition (6.119),

$$\tilde{F}(Dev\hat{\mathbf{S}}) = \tilde{F}(\mathbf{R}^t(Dev\hat{\mathbf{S}})\mathbf{R}), \tag{7.9}$$

where, in the case of isotropy, $\mathbf{R}$ is an arbitrary orthogonal tensor.

We have argued, in Section 6.5, that the yield function should be approximated by a quadratic function of $Dev\hat{\mathbf{S}}$, for consistency with other assumptions made in the course of setting up the theory. With reference to the Appendix to Section 6.1, the most general such function in the case of isotropy is a linear combination of $tr(Dev\hat{\mathbf{S}})$, $(tr(Dev\hat{\mathbf{S}}))^2$, and $tr((Dev\hat{\mathbf{S}})^2) = \left|Dev\hat{\mathbf{S}}\right|^2$, the first two of which vanish identically. The most general yield function of the required kind such that the yield surface $F = 0$ partitions stress space into regions defined by $F > 0$ and $F < 0$, in which the latter contains the stress-free state, is then of the form

$$\tilde{F}(Dev\hat{\mathbf{S}}) = \tfrac{1}{2}\left|Dev\hat{\mathbf{S}}\right|^2 - k^2. \tag{7.10}$$

This is the famous yield function proposed by von Mises and discussed in the books by Hill and Prager, among many others. The present derivation, based on material symmetry arguments, promotes understanding of its position in the context of the overall theory. We have derived von Mises' function as the most general yield function compatible with isotropic material symmetry, with pressure insensitivity, and with our further requirement of a quadratic dependence on $\hat{\mathbf{S}}$, the latter to ensure consistency with our similar approximation of the strain-energy function. Aficionados will object that our formulation therefore excludes the equally famous Tresca yield function from consideration. This is also compatible with isotropy and pressure insensitivity, but not with our further consistency requirement. However, on p. 21 of Hill's treatise, we find the statement "For most metals von Mises' criterion fits the data more closely than Tresca's." This conclusion, in turn, is based on the results of careful experiments conducted by Tayor and Quinney—see their classic paper, entitled "The Plastic Distortion of Metals," reproduced in G. I. Taylor's *Collected Works*. More recently, a critique of Tresca's criterion, and a strong preference for von Mises', has been offered in the book by Christensen. Thus, we consider our restriction to von Mises' function to be entirely appropriate.

Because the linear space of symmetric tensors may be decomposed as the direct sum of the five-dimensional space of deviatoric tensors and the one-dimensional space of spherical tensors, it follows that the yield surface, defined by $F = 0$, is a cylinder in six-dimensional $\hat{\mathbf{S}}$ -space, with axis $\mathbf{I}$ and radius $\sqrt{2}k$. The elastic range, defined by $F < 0$, is trivially convex. Here k is the yield stress in shear. That is, if the state of stress is a pure shear of the form

$$\hat{\mathbf{S}} = S(\mathbf{i} \otimes \mathbf{j} + \mathbf{j} \otimes \mathbf{i}), \tag{7.11}$$

with $\mathbf{i}$ and $\mathbf{j}$ orthonormal, then $\left|Dev\hat{\mathbf{S}}\right|^2 = 2S^2$ and the onset of yield occurs when $|S| = k$. Here k may be a fixed constant, corresponding to *perfect* plasticity, or may depend on appropriate variables that characterize the manner in which the state of the material evolves with plastic flow. The latter pertains to *work hardening*, the modeling of which is the central open problem of the phenomenological theory of plasticity. Some tentative ideas concerning such modeling are discussed in Chapter 9.

To generate the flow rule for the plastic deformation, we proceed as in (6.117), obtaining

$$\begin{aligned} F_{\hat{\mathbf{S}}} \cdot \hat{\mathbf{S}}' &= F' = \tilde{F}' = \tfrac{1}{2}(Dev\hat{\mathbf{S}} \cdot Dev\hat{\mathbf{S}})' \\ &= Dev\hat{\mathbf{S}} \cdot (Dev\hat{\mathbf{S}})' = Dev\hat{\mathbf{S}} \cdot Dev\hat{\mathbf{S}}' = Dev\hat{\mathbf{S}} \cdot \hat{\mathbf{S}}', \end{aligned} \tag{7.12}$$

and hence

$$F_{\hat{\mathbf{S}}} = Dev\hat{\mathbf{S}}. \tag{7.13}$$

This yields $\hat{\mathbf{S}} \cdot F_{\hat{\mathbf{S}}} = \left|Dev\hat{\mathbf{S}}\right|^2$, which conforms to the dissipation inequality (6.113) whenever the yield condition $F = 0$ is satisfied. Finally, (7.8) delivers the flow rule

$$\dot{\mathbf{G}}\mathbf{G}^{-1} = \lambda Dev\hat{\mathbf{S}}. \tag{7.14}$$

Problem 7.3. With reference to Problem 6.9, derive an expression for the yield function $F(\hat{\mathbf{S}})$ in the case of cubic symmetry. Assume this to be a quadratic function of $Dev\hat{\mathbf{S}}$, and derive an expression for $F_{\hat{\mathbf{S}}}$. State the form of the flow rule (6.105) in this case, together with appropriate restrictions on the plastic spin arising from considerations of material symmetry. Note that it is necessary to provide a constitutive prescription for the plastic spin. You need not do so here, but it would be worthwhile to research this issue independently. See, for example, the paper by Edmiston et al.

7.3 The classical theory for isotropic rigid-plastic materials

The elastic strain is invariably small in the case of rate-independent response or under low strain-rate conditions because it is then bounded by the diameter of the elastic range. If the overall strain is nevertheless large, then the main contribution to the strain comes from plastic deformation. In this case it is appropriate to consider the idealization of zero elastic strain, which entails the restriction $\mathbf{H}^t\mathbf{H} = \mathbf{I}$. The elastic deformation is therefore a rotation field, which we denote by $\mathbf{Q}$. The stress $\hat{\mathbf{S}}$ is then effectively a symmetric Lagrange-multiplier tensor associated with the constraint of vanishing elastic strain. It is constitutively indeterminate, as is the stress in a rigid body subject to the constraint $\mathbf{F}^t\mathbf{F} = \mathbf{I}$. In the present context it is subject only to the yield condition $F = 0$ and the balance laws, but is otherwise arbitrary. Further, $\mathcal{J}_H = 1$ and the relation $\mathcal{J}_H\mathbf{T} = \mathbf{H}\hat{\mathbf{S}}\mathbf{H}^t$ between the Cauchy stress $\mathbf{T}$ and Piola–Kirchhoff stress $\hat{\mathbf{S}}$, which may be inferred from (6.9) and (6.11), becomes

$$\hat{\mathbf{S}} = \mathbf{Q}^t\mathbf{T}\mathbf{Q}. \tag{7.15}$$

Therefore,

$$Dev\hat{\mathbf{S}} = Dev(\mathbf{Q}^t\mathbf{T}\mathbf{Q}) = \mathbf{Q}^t\mathbf{T}\mathbf{Q} - \tfrac{1}{3}tr(\mathbf{Q}^t\mathbf{T}\mathbf{Q})\mathbf{I} = \mathbf{Q}^t\boldsymbol{\tau}\mathbf{Q}, \tag{7.16}$$

where

$$\boldsymbol{\tau} = dev\mathbf{T} \equiv \mathbf{T} - \tfrac{1}{3}(tr\mathbf{T})\mathbf{i} \tag{7.17}$$

is the deviatoric part of the Cauchy stress in which $\mathbf{i}$ is the spatial identity. This differs from the deviatoric operator *Dev* defined in (6.115), based on the referential identity. The distinction is necessitated by the fact that the plastic deformation $\mathbf{G}$ is insensitive to superposed rigid-body motions, implying that $\kappa_i(p)$, like $T_{\mathcal{K}}$, is similarly insensitive. It is appropriate, then, to adopt $\mathbf{I}$ as the identity for both vector spaces. On the other hand, $\mathbf{T}$ preserves $T_{\mathcal{K}_t}$, so that (7.17) makes sense with the identity interpreted as the spatial identity. In contrast, $\mathbf{Q}$ behaves like the shifter $\mathbf{1}$, mapping $\kappa_i(p)$ to $T_{\mathcal{K}_t}$. Thus, $\mathbf{Q}^t$ maps the latter to the former. This implies that $\mathbf{Q}^t\mathbf{Q} = \mathbf{I}$ and $\mathbf{Q}\mathbf{Q}^t = \mathbf{i}$. To make this explicit we introduce a rotation $\bar{\mathbf{Q}}$, mapping $\kappa_i(p)$ to itself, such that $\mathbf{Q} = \mathbf{1}\bar{\mathbf{Q}}$. This makes sense because the composition of rotations is a rotation, and, as we have seen in Section 2.3, the shifter is a rotation. We then have

$$\tilde{F}(Dev\hat{\mathbf{S}}) = \tilde{F}(\mathbf{Q}^t\boldsymbol{\tau}\mathbf{Q}) = \tilde{F}(\bar{\mathbf{Q}}^t\mathbf{1}^t\boldsymbol{\tau}\mathbf{1}\bar{\mathbf{Q}}). \tag{7.18}$$

Now, $\bar{\mathbf{Q}}$ is a rotation of the same kind as the rotation $\mathbf{R}$ appearing in the symmetry condition (7.9); that is, $\mathbf{R}$, like $\bar{\mathbf{Q}}$, rotates vectors in $\kappa_i(p)$ to vectors in $\kappa_i(p)$. Accordingly, because $\mathbf{R}$ is otherwise arbitrary in the case of isotropy, we may invoke (7.9), with $\mathbf{R} = \bar{\mathbf{Q}}$, to conclude that $\tilde{F}(\bar{\mathbf{Q}}^t\mathbf{1}^t\boldsymbol{\tau}\mathbf{1}\bar{\mathbf{Q}}) = \tilde{F}(\mathbf{1}^t\boldsymbol{\tau}\mathbf{1})$, and hence that

$$\tilde{F}(Dev\hat{\mathbf{S}}) = \tilde{F}(\mathbf{1}^t\boldsymbol{\tau}\mathbf{1}). \tag{7.19}$$

This is meaningful insofar as $\mathbf{1}^t\boldsymbol{\tau}\mathbf{1}$ is a tensor of the same kind as $Dev\hat{\mathbf{S}}$; i.e., both preserve $\kappa_i(p)$. Substituting the von Mises form and using $\mathbf{1}\mathbf{1}^t = \mathbf{i}$, we have

$$\begin{aligned} \tilde{F}(Dev\hat{\mathbf{S}}) &= \tfrac{1}{2}tr(\mathbf{1}^t\boldsymbol{\tau}\mathbf{1}\mathbf{1}^t\boldsymbol{\tau}\mathbf{1}) - k^2 = \tfrac{1}{2}tr(\mathbf{1}^t\boldsymbol{\tau}^2\mathbf{1}) - k^2 \\ &= \tfrac{1}{2}tr(\boldsymbol{\tau}^2\mathbf{1}\mathbf{1}^t) - k^2 = \tfrac{1}{2}tr(\boldsymbol{\tau}^2) - k^2. \end{aligned} \tag{7.20}$$

Thus, $F(\hat{\mathbf{S}}) = \tilde{F}(\boldsymbol{\tau})$, where

$$\tilde{F}(\boldsymbol{\tau}) = \tfrac{1}{2}\,|\boldsymbol{\tau}|^2 - k^2. \tag{7.21}$$

This is the conventional form of von Mises' yield function found in the literature. It emerges from the formalism presented here when the elastic strain vanishes.

Using (7.16), we recast the flow rule (7.14) as

$$\mathbf{Q}(\dot{\mathbf{G}}\mathbf{G}^{-1})\mathbf{Q}^t = \lambda\boldsymbol{\tau}. \tag{7.22}$$

This may be reduced to the classical form via the decomposition

$$\mathbf{L} = \mathbf{D} + \mathbf{W} \tag{7.23}$$

of the spatial velocity gradient $\mathbf{L} = grad\mathbf{v}$ into the sum of the symmetric straining tensor $\mathbf{D}$ and the skew vorticity tensor $\mathbf{W}$. Then with $\mathbf{L} = \dot{\mathbf{F}}\mathbf{F}^{-1}$ and (5.3) we find, in the present specialization to $\mathbf{H} = \mathbf{Q}$, that

$$\mathbf{L} = \dot{\mathbf{Q}}\mathbf{Q}^t + \mathbf{Q}(\dot{\mathbf{G}}\mathbf{G}^{-1})\mathbf{Q}^t, \tag{7.24}$$

in which the first term is skew whereas the second, according to (7.22), is symmetric. The uniqueness of the decomposition then yields $\mathbf{W} = \dot{\mathbf{Q}}\mathbf{Q}^t$, together with $\mathbf{D} = \mathbf{Q}(\dot{\mathbf{G}}\mathbf{G}^{-1})\mathbf{Q}^t$, and hence the classical flow rule

$$\mathbf{D} = \lambda\boldsymbol{\tau}; \quad \lambda \geq 0, \tag{7.25}$$

proposed, albeit in a different form, of course, by St. Venant, in 1870, and Lévy, in 1871.

The Cauchy stress is given by

$$\mathbf{T} = \boldsymbol{\tau} - p\mathbf{i}, \tag{7.26}$$

where p is the pressure field. This pressure field is associated with the incompressibility of the material, i.e., with the constraint $tr\mathbf{D} = 0$, which follows from (7.25). It is not related to the constitutive response of the material, but is instead an additional unknown to be obtained by solving the equations of motion, augmented by the constraint equation together with any subsidiary conditions. Chadwick's book includes a discussion of the pressure associated with incompressibility, and of similar modifications to the expression for the stress when alternative constraints are operative.

Equation (7.25) is the fundamental equation of the classical theory, which predates the modern theory for finite elastic–plastic deformations, as discussed here, by at least a century. Its derivation in the framework of the modern theory, based on material symmetry restrictions pertaining to the concept of an undistorted state, is significant insofar as it promotes confidence in both the classical and modern theories. It is remarkable that the original form, deduced without the benefit of the tools of modern continuum mechanics, emerges in such a straightforward manner from the modern theory.

Indeed, prior to the advent of the modern theory, the classical theory was regarded as a separate model, seemingly unrelated to the theory for elastic–plastic response and without a clear connection to the notion of isotropic material symmetry as espoused in continuum mechanics. To elaborate, we take the inner product of (7.25) with itself to find λ, and, recalling that this equation is operative when $\tilde{F}(\boldsymbol{\tau}) = 0$, obtain $\boldsymbol{\tau} = \hat{\boldsymbol{\tau}}(\mathbf{D})$, where

$$\hat{\boldsymbol{\tau}}(\mathbf{D}) = \tfrac{\sqrt{2}k}{|\mathbf{D}|}\mathbf{D}. \tag{7.27}$$

This is an *isotropic function*; that is, $\hat{\boldsymbol{\tau}}(\mathbf{QDQ}^t) = \mathbf{Q}\hat{\boldsymbol{\tau}}(\mathbf{D})\mathbf{Q}^t$ for any (spatial) rotation $\mathbf{Q}$. However, this feature is a consequence of frame invariance, or invariance of $\hat{\boldsymbol{\tau}}(\mathbf{D})$ under superposed rigid-body motions. It is not related to the notion of isotropic material symmetry, except to the extent indicated in our derivation of (7.25). The claim, prevalent in the classical literature, to the effect that (7.25) is a model of isotropic solids, is thus difficult to reconcile with the modern notion of isotropy as discussed in the foregoing, and hence a potential source of confusion which we have endeavored here to clarify. It seems reasonable to view this difficulty as one of the reasons why plasticity theory received so little attention from the main proponents of the twentieth-century renaissance of continuum mechanics. In fact, from the modern perspective (7.27) is more appropriately interpreted as a model, operative when $|\boldsymbol{\tau}| = \sqrt{2}k$, of a non-Newtonian fluid, with a variable viscosity η given by

$$2\eta = \sqrt{2}k/\,|\mathbf{D}|\,. \tag{7.28}$$

Otherwise, i.e., when $|\boldsymbol{\tau}| < \sqrt{2}k$, we have $\mathbf{D} = \mathbf{0}$, because $\dot{\mathbf{G}}$ vanishes when $F < 0$. It is well known—see the books by Chadwick and Gurtin, for example—that in this case the deformation $\boldsymbol{\chi}$ is rigid; hence, the name *rigid-plastic material* attached to this model.

Problem 7.4. A popular yield function for isotropic soils is the Drucker–Prager function

$$F(\hat{\mathbf{S}}) = \tfrac{1}{2}\left|Dev\hat{\mathbf{S}}\right|^2 - [\alpha(tr\hat{\mathbf{S}}) + k]^2,$$

where α and k are material constants. What is the corresponding flow rule for $\dot{\mathbf{G}}\mathbf{G}^{-1}$? Reduce this to an equation for the straining tensor $\mathbf{D}$ in the case of rigid-plastic response. Is the material compressible or incompressible? (For a thorough discussion of soil plasticity, consult the book by Yu.)

7.4 Bingham's model of viscoplasticity

We will study the rigid-plastic model in some detail in Section 7.5, but, to expand on our remarks concerning the interpretation of (7.27) as a model of viscous fluids, we first pause to consider a model of viscoplastic materials proposed, in 1922, by Bingham. Prager's book contains a thorough discussion. This model is obtained simply by adding a conventional viscous term to the right-hand side of (7.27),

$$\boldsymbol{\tau} = (2\eta + \tfrac{\sqrt{2}k}{|\mathbf{D}|})\mathbf{D}, \qquad (7.29)$$

where η is now a fixed positive constant, and the Cauchy stress is again of the form (7.26). If we were to include only 2η within the parentheses, then we would have the classical Newtonian fluid in which η is the constant shear viscosity. Accordingly, (7.29) combines the effects of viscosity and plasticity, and thus furnishes a prototypical model of *viscoplastic* materials. Taking its norm, we have

$$|\boldsymbol{\tau}| = 2\eta\,|\mathbf{D}| + \sqrt{2}k. \qquad (7.30)$$

Thus, if the motion is non-rigid, that is, if $\mathbf{D} \neq \mathbf{0}$, then the stress necessarily lies *outside* the yield surface, i.e., $|\boldsymbol{\tau}| \geq \sqrt{2}k$. If $|\boldsymbol{\tau}| < \sqrt{2}k$, then this model is not applicable; we assume in this case that $\mathbf{D}$ vanishes, as before, and hence that the motion is rigid. Equation (7.29) thus models *rigid-viscoplastic* materials.

Further, observe that $|\boldsymbol{\tau}|$ is an increasing function of $|\mathbf{D}|$ whenever $|\boldsymbol{\tau}| \geq \sqrt{2}k$, whereas $|\boldsymbol{\tau}| \to \sqrt{2}k$ as η and/or $|\mathbf{D}|$ approach zero. Thus, (7.29) exhibits *rate-sensitivity*—a feature of the high-temperature response of real metals—and reverts to the conventional rate-independent model in the slow-deformation limit. It also furnishes a fairly realistic rheological model of wet paint.

The viscoplastic theory plays an important role in the solution of dynamical problems, extensive discussions of which may be found in the books by Cristescu, Kachanov, and Lubliner.

7.4.1 Example: Steady channel flow

To illustrate the main features of Bingham's model we apply it to the simple problem of steady rectilinear channel flow. Our solution is adapted from that given in Prager's book. Thus, consider the steady flow of a Bingham material in a channel: a region defined, in terms of Cartesian coordinates x, y, z, by $-\infty < x < \infty$, $0 < y < h$, and $-\infty < z < \infty$. We identify these coordinates with the projections of the position $\mathbf{y} \in \kappa_t$ of a material point (4.1) onto the elements of a fixed orthonormal basis $\{\mathbf{i}, \mathbf{j}, \mathbf{k}\}$; thus, $\mathbf{y} = x\mathbf{i} + y\mathbf{j} + z\mathbf{k}$.

We use the theory to investigate the possibility of the existence of a velocity field of the form

$$\mathbf{v} = v(y)\mathbf{i}. \tag{7.31}$$

The spatial velocity gradient $\mathbf{L}$ follows from

$$\mathbf{L}d\mathbf{y} = d\mathbf{v} = v'(y)\mathbf{i}dy = v'(y)\mathbf{i}(\mathbf{j} \cdot d\mathbf{y}) = \{v'(y)\mathbf{i} \otimes \mathbf{j}\}d\mathbf{y}, \tag{7.32}$$

giving $\mathbf{L} = v'(y)\mathbf{i} \otimes \mathbf{j}$ and

$$\mathbf{D} = \tfrac{1}{2}v'(y)(\mathbf{i} \otimes \mathbf{j} + \mathbf{j} \otimes \mathbf{i}), \tag{7.33}$$

and hence $tr\mathbf{D} = 0$, so that the assumed flow is kinematically possible in our idealized viscoplastic material. Further, as $\mathbf{L}\mathbf{v} = \mathbf{0}$ and because $\mathbf{v}_t$ vanishes, that is, because $\mathbf{v}$ is steady in the sense that it is independent of t when expressed as a function of $\mathbf{y}$ and t, we have $\dot{\mathbf{v}} = \mathbf{0}$, where $\dot{\mathbf{v}} = \mathbf{v}_t + \mathbf{L}\mathbf{v}$ is the material acceleration. Assuming negligible body force, (1.32) then requires that

$$div\mathbf{T} = \mathbf{0}, \tag{7.34}$$

with $\mathbf{T}$ given by (7.26) and (7.29), provided that $|\boldsymbol{\tau}| \geq \sqrt{2}k$. Later, we will need to interpret this inequality in terms of the function $v(y)$. Preliminary to this, we compute

$$\mathbf{D}^2 = \tfrac{1}{4}(v')^2(\mathbf{i} \otimes \mathbf{i} + \mathbf{j} \otimes \mathbf{j}) \tag{7.35}$$

and $|\mathbf{D}|^2 = tr(\mathbf{D}^2) = \tfrac{1}{2}(v')^2$. Thus, $|\mathbf{D}| = |v'(y)|/\sqrt{2}$, and (7.29) yields

$$\boldsymbol{\tau} = \tau(y)(\mathbf{i} \otimes \mathbf{j} + \mathbf{j} \otimes \mathbf{i}), \quad \text{where} \quad \tau(y) = \eta v'(y) + kv'(y)/|v'(y)|. \tag{7.36}$$

Substituting this into (7.26), we find that (7.34) reduces to

$$gradp = \tau'(y)\mathbf{i}, \tag{7.37}$$

which implies that $\partial p/\partial y = \partial p/\partial z = 0$, and $gradp = p'(x)\mathbf{i}$, where $p'(x) = \tau'(y)$. Thus, $p(x)$ is the linear function of x given by $p(x) = x\tau'(y) + p_0$, where p_0 is a constant, and since this is independent of y, we must have $x\tau''(y) = 0$; hence,

$$\tau'(y) = -C \quad \text{and} \quad p(x) = p_0 - Cx, \tag{7.38}$$

where C is another constant. We take $C > 0$, corresponding to a negative pressure gradient along the x-axis.

We assume the flow to be attached to stationary rigid walls at $y = 0, h$, where the conventional no-slip conditions hold, and we seek a velocity profile that is symmetric about the plane $y = h/2$. This means that $v'(y)$ vanishes at $y = h/2$. Because of symmetry we further suppose that $\tau(h/2) = 0$, this following automatically in the case of purely viscous response ($k = 0$). Thus,

$$\tau(y) = \tfrac{1}{2}C(h - 2y). \tag{7.39}$$

For the purely viscous fluid this gives $v'(y) = \tau(y)/\eta$, and hence the classical parabolic profile

$$v(y) = \tfrac{1}{2\eta}Cy(h - 2y), \tag{7.40}$$

satisfying the no-slip conditions $v(0) = v(h) = 0$. However, in the viscoplastic case we require $|\boldsymbol{\tau}| \geq \sqrt{2}k$, where, from $(7.36)_1$, $|\boldsymbol{\tau}| = \sqrt{2}\,|\tau|$. Thus, viscoplastic flow is possible only if

$$|\tau(y)| \geq k. \tag{7.41}$$

Consider the region where $\tau(y) > 0$. Then we have flow wherever

$$\tau(y) \geq k, \tag{7.42}$$

i.e., in the region

$$y \leq \bar{y}, \quad \text{where} \quad \bar{y} = \tfrac{h}{2} - \tfrac{k}{C}, \tag{7.43}$$

which makes sense provided that $\bar{y} > 0$, i.e., provided that the pressure gradient C satisfies

$$C > \tfrac{2k}{h}. \tag{7.44}$$

Evidently, then, these conditions require that $y < h/2$. In view of the no-slip condition $v(0) = 0$, we assume that $v'(y) \geq 0$ in this region, and conclude, from $(7.36)_2$, that

$$v'(y) = \tfrac{1}{\eta}(\tau - k). \tag{7.45}$$

The assumption about $v'(y)$ is then consistent with the requirement (7.42). Substituting (7.39) and imposing the no-slip condition, we arrive at the velocity field

$$v(y) = \tfrac{C}{2\eta}\left[y(h - y) - \tfrac{2k}{C}y\right]; \quad 0 \leq y \leq \bar{y}. \tag{7.46}$$

In the region $\bar{y} < y \leq h/2$ we still have $\tau(y) \geq 0$, but $|\tau(y)| < k$. Thus, $\mathbf{D} = \mathbf{0}$ and hence $v'(y) = 0$, implying that that the velocity is uniform in this region:

$$v(y) = v(\bar{y}); \quad \bar{y}<y \leq h/2. \tag{7.47}$$

This region moves as a rigid plug, provided, of course, that (7.44) is satisfied. Otherwise, the no-slip condition implies that there is no flow at all. Accordingly, flow initiates when the pressure gradient attains the critical value $C = 2k/h$. Naturally the solution in the region $h/2 < y < h$ follows from that already derived by invoking symmetry.

Problem 7.5. Position in a cylindrical polar-coordinate system $\{r, \varphi, z\}$ is given by

$$\mathbf{y} = r\mathbf{e}_r(\varphi) + z\mathbf{k},$$

where $r = \sqrt{x^2 + y^2}$, $\tan\varphi = y/x$, $\mathbf{e}_r(\varphi) = \cos\varphi\mathbf{i}_1 + \sin\varphi\mathbf{i}_2$, and $\mathbf{k} = \mathbf{i}_3$. Using the basis $\{\mathbf{e}_r, \mathbf{e}_\varphi, \mathbf{k}\}$, where $\mathbf{e}_\varphi(\varphi) = \mathbf{k} \times \mathbf{e}_r$, we can write the Cauchy stress tensor in the form

$$\begin{aligned}\mathbf{T} \;=\;& T_{rr}\mathbf{e}_r \otimes \mathbf{e}_r + T_{\varphi\varphi}\mathbf{e}_\varphi \otimes \mathbf{e}_\varphi + T_{zz}\mathbf{k} \otimes \mathbf{k} + T_{r\varphi}(\mathbf{e}_r \otimes \mathbf{e}_\varphi + \mathbf{e}_\varphi \otimes \mathbf{e}_r) \\ &+ T_{rz}(\mathbf{e}_r \otimes \mathbf{k} + \mathbf{k} \otimes \mathbf{e}_r) + T_{\varphi z}(\mathbf{e}_\varphi \otimes \mathbf{k} + \mathbf{k} \otimes \mathbf{e}_\varphi).\end{aligned}$$

Assuming all components T_{rr}, $T_{r\varphi}$, etc., to depend only on r and φ, use the methods of Chapter 3, or otherwise, to show that

$$\begin{aligned}div\mathbf{T} \;=\;& \left[\frac{\partial}{\partial r}T_{rr} + \frac{1}{r}\left(T_{rr} - T_{\varphi\varphi} + \frac{\partial}{\partial\varphi}T_{r\varphi}\right)\right]\mathbf{e}_r \\ &+ \left[\frac{\partial}{\partial r}T_{r\varphi} + \frac{1}{r}\left(2T_{r\varphi} + \frac{\partial}{\partial\varphi}T_{\varphi\varphi}\right)\right]\mathbf{e}_\varphi \\ &+ \left[\frac{\partial}{\partial r}T_{rz} + \frac{1}{r}\left(T_{rz} + \frac{\partial}{\partial\varphi}T_{\varphi z}\right)\right]\mathbf{k}.\end{aligned}$$

Problem 7.6. Use Bingham's model to solve the problem of anti-plane viscoplastic flow in a circular pipe aligned with the z-axis. Assume a velocity field of the form $\mathbf{v} = w(r)\mathbf{k}$, where $r = \sqrt{x^2 + y^2}$ is the radius in a system of cylindrical polar coordinates. Suppose the material occupies the annular region $(0 <)a \leq r \leq b$ and that $w(b) = 0$ and $w(a) = W$. Assume that the axial pressure gradient vanishes and find the velocity field $w(r)$.

7.5 Plane strain of rigid-perfectly plastic materials: Slip-line theory

The rate-independent rigid-perfectly plastic model of Section 7.3 takes a particularly interesting form when specialized to plane-strain conditions. The theory, thus specialized, lends itself to the solution of many practically important problems and is thus the subject of an extensive literature. For this reason we devote a fair amount of attention to the plane-strain theory here.

Thus, we confine attention to velocity fields of the form $\mathbf{v} = v_\alpha(y^1, y^2)\mathbf{i}_\alpha$, where Greek indices are summed, here and henceforth, from 1 to 2, $\{y^1, y^2\}$ is a system of plane Cartesian coordinates, and the $\mathbf{i}_\alpha$ are fixed unit vectors defined by (3.12). The latter equation implies, as is well known, that there is no distinction between co- and contravariance when using such coordinates. Accordingly, we will adopt the common convention of simply lowering all indices. Then, $\mathbf{D} = D_{\alpha\beta}\mathbf{i}_\alpha \otimes \mathbf{i}_\beta$, where $2D_{\alpha\beta} = v_{\alpha,\beta} + v_{\beta,\alpha}$, with $v_{\alpha,\beta} = \partial v_\alpha / \partial y_\beta$, and Eq. (7.25) yields $\boldsymbol{\tau} = \tau_{\alpha\beta}\mathbf{i}_\alpha \otimes \mathbf{i}_\beta$, implying that $p = -T_{33}$. The pressure field is equal to the confining stress required to maintain the plane-strain condition. Using $3p = -tr\mathbf{T}$, we conclude that $p = -\frac{1}{2}T_{\alpha\alpha}$.

7.5.1 Stress, equilibrium

The yield criterion $F = 0$ reduces, with (7.21), to

$$\begin{aligned} 2k^2 &= \tau_{\alpha\beta}\tau_{\alpha\beta} = (T_{\alpha\beta} + p\delta_{\alpha\beta})(T_{\alpha\beta} + p\delta_{\alpha\beta}) \\ &= T_{\alpha\beta}T_{\alpha\beta} + 2pT_{\alpha\alpha} + p^2\delta_{\alpha\alpha} \\ &= T_{\alpha\beta}T_{\alpha\beta} - 2p^2 = T_{\alpha\beta}T_{\alpha\beta} - \tfrac{1}{2}(T_{\alpha\alpha})^2, \end{aligned} \tag{7.48}$$

where $\delta_{\alpha\beta}$ is the Kronecker delta, or, more simply, to

$$(T_{11} - T_{22})^2 + 4T_{12}^2 = 4k^2. \tag{7.49}$$

Problem 7.7. Show that the principal stresses are

$$T_1 = -p + k, \quad T_2 = -p - k \quad \text{and} \quad T_3 = -p.$$

It follows that

$$\mathbf{T} = -p\mathbf{i} + k(\mathbf{u}_1 \otimes \mathbf{u}_1 - \mathbf{u}_2 \otimes \mathbf{u}_2), \tag{7.50}$$

where $\{\mathbf{u}_i\}$, with $\mathbf{u}_3 = \mathbf{i}_3$, are the orthonormal principal stress axes. Let $\mathbf{t}$ and $\mathbf{s}$ be plane orthonormal vector fields such that

$$\mathbf{u}_1 = \tfrac{\sqrt{2}}{2}(\mathbf{s} + \mathbf{t}), \quad \mathbf{u}_2 = \tfrac{\sqrt{2}}{2}(\mathbf{s} - \mathbf{t}). \tag{7.51}$$

Then,

$$\mathbf{T} = -p\mathbf{i} + \boldsymbol{\tau}, \quad \text{where} \quad \boldsymbol{\tau} = k(\mathbf{t} \otimes \mathbf{s} + \mathbf{s} \otimes \mathbf{t}), \tag{7.52}$$

and this implies that k is the shear stress resolved on the $\mathbf{s}, \mathbf{t}$ -axes.

Problem 7.8. If $\mathbf{a}(\mathbf{y})$ and $\mathbf{b}(\mathbf{y})$ are vector fields, show that $div(\mathbf{a} \otimes \mathbf{b}) = (grad\mathbf{a})\mathbf{b} + (div\mathbf{b})\mathbf{a}$, where *grad* and *div* are the gradient and divergence operators based on position $\mathbf{y}$.

Accordingly, in a perfectly plastic material ($k = const.$), equilibrium without body force, i.e., $div\mathbf{T} = \mathbf{0}$, requires that

$$grad(p/k) = (grad\mathbf{t})\mathbf{s} + (div\mathbf{s})\mathbf{t} + (grad\mathbf{s})\mathbf{t} + (div\mathbf{t})\mathbf{s}. \tag{7.53}$$

If we introduce a two-dimensional scalar field, $\theta(\mathbf{y})$, such that

$$\mathbf{t} = \cos\theta\mathbf{i}_1 + \sin\theta\mathbf{i}_2, \quad \mathbf{s} = -\sin\theta\mathbf{i}_1 + \cos\theta\mathbf{i}_2, \tag{7.54}$$

then

$$d\mathbf{t} = \mathbf{s}d\theta \quad \text{and} \quad d\mathbf{s} = -\mathbf{t}d\theta, \tag{7.55}$$

and thus, with $d\theta = grad\theta \cdot d\mathbf{y}$,

$$grad\mathbf{t} = \mathbf{s} \otimes grad\theta, \quad grad\mathbf{s} = -\mathbf{t} \otimes grad\theta, \tag{7.56}$$

and therefore, with the divergence defined by (3.71),

$$div\mathbf{t} = \mathbf{s} \cdot grad\theta \quad \text{and} \quad div\mathbf{s} = -\mathbf{t} \cdot grad\theta. \tag{7.57}$$

We thus reduce (7.53) to

$$grad(p/2k) = (\mathbf{s} \cdot grad\theta)\mathbf{s} - (\mathbf{t} \cdot grad\theta)\mathbf{t}, \tag{7.58}$$

which is equivalent to the two equations

$$\mathbf{t} \cdot grad(p/2k + \theta) = 0 \quad \text{and} \quad \mathbf{s} \cdot grad(p/2k - \theta) = 0, \tag{7.59}$$

due to Prandtl and Hencky.

These require that $p/2k \pm \theta$ take constant values on the trajectories defined by $dy_2/dy_1 = \tan\theta, -\cot\theta$, respectively. The latter are the *characteristic curves* of a *hyperbolic* system of partial differential equations for the stress fields p and θ. Remarkably, the stress is *statically determinate* ; that is, granted suitable boundary conditions, it can be determined directly from the equilibrium equations alone without knowledge of the motion. These striking features of the theory of perfectly plastic materials contrast sharply with situation encountered in the theory of elasticity, for example, where the equilibrium equations for the displacement field are typically of *elliptic* type.

7.5.2 Velocity field

Naturally, there must be a velocity field to accompany the stress field, be it statically determinate or not. To obtain this, it proves advantageous to decompose the velocity field in the variable $\{\mathbf{s}, \mathbf{t}\}$ basis. Thus,

$$\mathbf{v} = v_t \mathbf{t} + v_s \mathbf{s}. \tag{7.60}$$

To compute the velocity gradient $\mathbf{L} = grad\mathbf{v}$, we use (7.55), obtaining

$$\begin{aligned} \mathbf{L}d\mathbf{y} &= d\mathbf{v} = dv_t \mathbf{t} + v_t \mathbf{s} d\theta + dv_s \mathbf{s} - v_s \mathbf{t} d\theta \\ &= (grad v_t \cdot d\mathbf{y})\mathbf{t} + v_t \mathbf{s}(grad\theta \cdot d\mathbf{y}) + (grad v_s \cdot d\mathbf{y})\mathbf{s} - v_s \mathbf{t}(grad\theta \cdot d\mathbf{y}). \end{aligned} \tag{7.61}$$

Thus,

$$\mathbf{L} = \mathbf{t} \otimes grad v_t + v_t \mathbf{s} \otimes grad\theta + \mathbf{s} \otimes grad v_s - v_s \mathbf{t} \otimes grad\theta, \tag{7.62}$$

and

$$\begin{aligned} 2\mathbf{D} &= \mathbf{t} \otimes grad v_t + grad v_t \otimes \mathbf{t} + \mathbf{s} \otimes grad v_s + grad v_s \otimes \mathbf{s} \\ &\quad + v_t(\mathbf{s} \otimes grad\theta + grad\theta \otimes \mathbf{s}) - v_s(\mathbf{t} \otimes grad\theta + grad\theta \otimes \mathbf{t}). \end{aligned} \tag{7.63}$$

Substituting into (7.25) and $(7.52)_2$, we derive

$$\mathbf{t} \cdot \mathbf{D}\mathbf{t} = 0 \quad \text{and} \quad \mathbf{s} \cdot \mathbf{D}\mathbf{s} = 0, \tag{7.64}$$

which together imply that the motion is isochoric, as required, with vanishing extension rates in the directions of $\mathbf{t}$ and $\mathbf{s}$.

Problem 7.9. Show that Eqs. (7.64) are equivalent to the pair

$$\mathbf{t} \cdot (grad v_t - v_s grad\theta) = 0 \quad \text{and} \quad \mathbf{s} \cdot (grad v_s + v_t grad\theta) = 0.$$

These are the celebrated Geiringer equations. If the stress field is known, they furnish linear partial differential equations for the velocity components v_t and v_s.

Remarkably, this theory allows for the existence of tangential discontinuities of the velocity field along the trajectories with unit tangents $\mathbf{s}$ and $\mathbf{t}$. To see that only tangential discontinuities are allowed, consider the local condition of conservation of mass at a surface of discontinuity. We derived the referential form of this condition, holding in κ, in Chapter 6—see (6.139). Here, however, we require its counterpart in κ_t. Rather than derive this condition anew, we refer to its derivation in Liu's book. The relevant condition, given by Liu's Eq. (2.32), is

$$v[\![\rho]\!] = [\![\rho \mathbf{v}]\!] \cdot \mathbf{n}, \tag{7.65}$$

where $[\![\cdot]\!]$ is the jump at the discontinuity surface with unit normal $\mathbf{n}$, and v is the normal speed of the surface in the direction of $\mathbf{n}$. Recalling that $\rho_\kappa = \rho \mathcal{J}_F$ and assuming ρ_κ to be continuous, we conclude, in the case of the isochoric motions ($J_F = 1$) that are possible in incompressible materials, that $[\rho]$ vanishes, and hence that $[\mathbf{v}] \cdot \mathbf{n} = 0$. Any velocity discontinuity must then be purely tangential to the discontinuity surface.

This implies that the normal velocity v_s (resp., v_t) is continuous across the curve with unit-tangent field $\mathbf{t}$ (resp., $\mathbf{s}$). The tangential velocity v_t (resp., v_s) may assume unequal values on either side of such a curve, and hence so too the tangential derivative $\mathbf{t} \cdot grad v_t$ (resp., $\mathbf{s} \cdot grad v_s$). Subtracting the limits of the relevant Geiringer equation on either side of the curve, we conclude that $\mathbf{t} \cdot grad[v_t]$ (resp., $\mathbf{s} \cdot grad[v_s]$) vanishes on the curve with normal $\mathbf{s}$ (resp., $\mathbf{t}$), so that the *slip* $[v_t]$ (resp., $[v_s]$), if non-zero, is uniform along it. This explains why the present model is referred to as *slip-line theory*.

The literature on this subject, relating in particular to the solution of problems involving metal forming and extrusion, is enormous. The books by Nadai, Hill, Kachanov, Johnson, and Haddow, Johnson and Mellor, and Lubliner, and the article by Geiringer describe further theory and many worked-out solutions. Numerical solutions are discussed in the article by Collins.

There is also a plane-strain slip-line theory for crystalline materials. This is not as nearly as well known as the theory for isotropic materials. It is substantially more complicated than the isotropic theory, and therefore not discussed here. However, an account of this subject, accessible to readers of this book, may by found in the paper by Gupta and Steigmann and the references cited therein.

We limit ourselves to a discussion of a few explicit solutions, both to illustrate some of the more unusual features of the theory and to whet the reader's appetite for the vast literature on the subject.

7.5.3 Cartesian form of the equations

Most of the literature is based on equations cast in terms of Cartesian coordinates.

Problem 7.10. Using (7.54), show that $\mathbf{T} = T_{\alpha\beta}\mathbf{i}_\alpha \otimes \mathbf{i}_\beta - p\mathbf{i}_3 \otimes \mathbf{i}_3$, where

$$T_{11} = -p - k\sin 2\theta, \quad T_{22} = -p + k\sin 2\theta, \quad \text{and} \quad T_{12} = T_{21} = k\cos 2\theta.$$

Equilibrium without body force, i.e., $T_{ij,j} = 0$, thus requires that $p_{,3} = 0$ and $T_{\alpha\beta,\beta} = 0$, the latter yielding

$$p_{,1} + 2k(\theta_{,1}\cos 2\theta + \theta_{,2}\sin 2\theta) = 0 \quad \text{and} \quad p_{,2} + 2k(\theta_{,1}\sin 2\theta - \theta_{,2}\cos 2\theta) = 0. \tag{7.66}$$

Note that if, at a particular point, we orient the basis $\{\mathbf{i}_\alpha\}$ such that $\mathbf{i}_1 = \mathbf{t}$ and $\mathbf{i}_2 = \mathbf{s}$, then θ vanishes at that point, and these equations reduce to

$$\begin{aligned} 0 &= p_{,1} + 2k\theta_{,1} = \mathbf{i}_1 \cdot grad(p + 2k\theta) = \mathbf{t} \cdot grad(p + 2k\theta), \quad \text{and, similarly,} \\ 0 &= p_{,2} - 2k\theta_{,2} = \mathbf{i}_2 \cdot grad(p - 2k\theta) = \mathbf{s} \cdot grad(p - 2k\theta), \end{aligned} \tag{7.67}$$

which, of course, are just the Prandtl–Hencky equations.

With the Cartesian form of the equations in hand, the obvious thing to do when searching for simple solutions is to assume that p or θ depends on just one of the coordinates. Let's consider the possibility that θ depends only on y_2; Eqs. (7.66) reduce to

$$p_{,1} = k(\cos 2\theta)' \quad \text{and} \quad p_{,2} = k(\sin 2\theta)', \tag{7.68}$$

where $(\cdot)' = d(\cdot)/dy_2$. Assuming p to be twice differentiable, we can eliminate it as follows:

$$k(\cos 2\theta)'' = p_{,12} = p_{,21} = k(\sin 2\theta)_{,21} = 0. \tag{7.69}$$

Then $\cos 2\theta$ is a linear function of y_2. One such solution, due to Prandtl, is

$$\cos 2\theta = -y_2/h, \quad \sin 2\theta = -\sqrt{1-(y_2/h)^2}, \tag{7.70}$$

where h is a positive constant. This is meaningful provided that $|y_2| \leq h$, and from (7.68) we have

$$p = -k(y_1/h) + f(y_2), \quad \text{where} \quad f' = k(\sin 2\theta)'. \tag{7.71}$$

Thus,

$$p = p_0 - k[y_1/h + \sqrt{1-(y_2/h)^2}], \tag{7.72}$$

where p_0 is a constant, and the equilibrium stress field follows from the solution to Problem 7.10. Our solution is adapted from Prager's version of Prandtl's solution.

To obtain the velocity field, we invoke (7.25) with

$$\boldsymbol{\tau} = k \sin 2\theta(\mathbf{i}_2 \otimes \mathbf{i}_2 - \mathbf{i}_1 \otimes \mathbf{i}_1) + k \cos 2\theta(\mathbf{i}_1 \otimes \mathbf{i}_2 + \mathbf{i}_2 \otimes \mathbf{i}_1). \tag{7.73}$$

Then,

$$\mathbf{D} = \alpha(\mathbf{i}_1 \otimes \mathbf{i}_1 - \mathbf{i}_2 \otimes \mathbf{i}_2) + \beta(\mathbf{i}_1 \otimes \mathbf{i}_2 + \mathbf{i}_2 \otimes \mathbf{i}_1), \tag{7.74}$$

where

$$\alpha = -\lambda k \sin 2\theta \quad \text{and} \quad \beta = \lambda k \cos 2\theta. \tag{7.75}$$

Because $\lambda \geq 0$ we then have $\lambda k = \sqrt{\alpha^2+\beta^2}$ and

$$\alpha/\sqrt{\alpha^2+\beta^2} = -\sin 2\theta, \quad \beta/\sqrt{\alpha^2+\beta^2} = \cos 2\theta, \tag{7.76}$$

specializing, in the present solution, to

$$\alpha = \sqrt{1-(y_2/h)^2}\sqrt{\alpha^2+\beta^2} \quad \text{and} \quad \beta = -(y_2/h)\sqrt{\alpha^2+\beta^2}. \tag{7.77}$$

It follows from the expression for **D** that the components of the velocity field satisfy

$$v_{1,1} = -v_{2,2} = \alpha \quad \text{and} \quad \tfrac{1}{2}(v_{1,2}+v_{2,1}) = \beta. \tag{7.78}$$

Accordingly, α and β are not arbitrary. Assuming the velocity components to be thrice differentiable, we derive the compatibility condition

$$\alpha_{,11} - \alpha_{,22} + 2\beta_{,12} = 0, \tag{7.79}$$

which, with reference to (7.77), is easily seen to be solved if

$$\sqrt{\alpha^2+\beta^2} = \tfrac{C}{h}\left(\sqrt{1-(y_2/h)^2}\right)^{-1}; \quad C = const. \tag{7.80}$$

For we then have

$$\alpha = \tfrac{C}{h} \quad \text{and} \quad \beta = -\tfrac{C}{h}\tfrac{y_2}{h}\left(\sqrt{1-(y_2/h)^2}\right)^{-1}, \tag{7.81}$$

a constant and a function of y_2 alone, respectively.

The first and second of Eqs. (7.78) integrate to give

$$v_1 = \tfrac{C}{h}y_1 + f(y_2) \quad \text{and} \quad v_2 = -\tfrac{C}{h}y_2 + g(y_1), \tag{7.82}$$

where f and g are functions to be determined. These yield $v_{1,2} = f'(y_2)$ and $v_{2,1} = g'(y_1)$, which combine with the third equality in (7.78) and the second in (7.81) to give $g''(y_1) = 0$ and

$$v_2 = -\tfrac{C}{h}y_2 + Dy_1 + E, \tag{7.83}$$

where D and E are constants. Then, $v_{1,2} = 2\beta - D$, yielding

$$v_1 = C[y_1/h + 2\sqrt{1-(y_2/h)^2}] - Dy_2 + F, \tag{7.84}$$

where F is another constant.

Clearly E and F are rigid-body translational velocities. Further, the terms involving D contribute a fixed tensor to the vorticity **W**, and therefore represent a rigid-body rotation. Suppressing the rigid-body terms, we thus have the velocity field

$$v_1 = C[y_1/h + 2\sqrt{1-(y_2/h)^2}], \quad v_2 = -Cy_2/h, \tag{7.85}$$

valid in the region defined by $|y_2| \leq h$.

Evidently $v_2 = \mp C$ at $y_2 = \pm h$, and so this solution simulates the squeezing, assuming $C > 0$, of a layer of thickness $2h$ between rigid walls approaching each other at the relative speed $2C$. The solution predicts that $T_{12} = \mp k$ at $y_2 = \pm h$ but does not satisfy the no-slip conditions $v_1 = 0$ there. We also note that $|D_{12}|\,(= |\beta|) \to \infty$ as $|y_2| \to h$. Thus, if the material is viscoplastic, then the viscosity cannot be neglected when y_2 is close to $\pm h$. In this case the present solution is valid outside a thin boundary layer adjoining the walls in which Bingham's model may be used to meet a no-slip condition. See Prager's book for a discussion of the relevant boundary-layer theory. Further discussion of the present solution is given in the books by Nadai and Hill, the latter including corrections to accommodate finite-width strips.

Corresponding slip lines may be determined, if desired, by using $(7.70)_1$ to obtain

$$dy_2/dy_1 = 2h(\sin 2\theta)d\theta/dy_1. \tag{7.86}$$

Then, on the trajectory with unit tangent **t**, where $dy_2/dy_1 = \tan\theta$, and on that with unit tangent **s**, where $dy_2/dy_1 = -\cot\theta$, we have

$$y_1 = h(\sin 2\theta + 2\theta) + const. \quad \text{and} \quad y_1 = h(\sin 2\theta - 2\theta) + const., \tag{7.87}$$

respectively. These, together with $(7.70)_1$, are the parametric equations of a pair of orthogonal cycloids.

7.5.4 Further theory for plane strain

To investigate the possibility of finding further solutions, we return to the Prandtl–Hencky form of the equations. It proves convenient for this purpose to retain Cartesian coordinates, but to ease the notation we label them $y_1 = x$ and $y_2 = y$ and reserve subscripts to identify partial derivatives. It is also convenient to introduce orthogonal curvilinear coordinates (α, β) to parametrize the slip lines (not to be confused with the use of the same symbols in the previous subsection). We arrange these such that β is constant on a trajectory with unit tangent **t**, and α is constant on one having unit tangent **s**. We refer to these trajectories as α-lines and β-lines, respectively. They satisfy the equations

$$(\tfrac{dy}{dx})_{|\beta} = \tan\theta \quad \text{and} \quad (\tfrac{dy}{dx})_{|\alpha} = -\cot\theta. \tag{7.88}$$

The Prandtl–Hencky equations (7.59) imply that $\pm p/2k + \theta$, respectively, assume constant values on these curves, and hence that

$$p/2k + \theta = \xi(\beta) \quad \text{and} \quad -p/2k + \theta = \eta(\alpha), \tag{7.89}$$

for some functions ξ and η. Similarly, the Geiringer equations require that

$$dv_t - v_s d\theta = 0 \quad \text{on} \quad \alpha\text{-lines, and} \quad dv_s + v_t d\theta = 0 \quad \text{on} \quad \beta\text{-lines.} \tag{7.90}$$

Using the chain rule, we have $d\xi = \xi_x dx + \xi_y dy = (\xi_x + \xi_y dy/dx)dx$. Then, because $d\xi = 0$ on an α-line,

$$\xi_x + (\tan\theta)\xi_y = 0. \tag{7.91}$$

In the same way, because $d\eta = 0$ on a β-line,

$$\eta_x - (\cot\theta)\eta_y = 0. \tag{7.92}$$

Noting, from (7.89), that

$$\theta = \tfrac{1}{2}(\xi + \eta), \tag{7.93}$$

these are seen to furnish a pair of quasilinear partial differential equations for the fields $\xi = \hat{\xi}(x, y)$ and $\eta = \hat{\eta}(x, y)$, solutions to which may be substituted into (7.89) to furnish the associated stress fields.

A standard approach to systems such as this is the *hodograph transformation*, whereby the roles of the dependent and independent variables are interchanged to arrive at a linear system for the latter, regarded as functions of the former. That is, we seek a system to be solved for the functions $x = \hat{x}(\xi, \eta)$ and $y = \hat{y}(\xi, \eta)$. To derive this, we invert the system

$$\begin{Bmatrix} d\xi \\ d\eta \end{Bmatrix} = \begin{pmatrix} \xi_x & \xi_y \\ \eta_x & \eta_y \end{pmatrix} \begin{Bmatrix} dx \\ dy \end{Bmatrix}, \tag{7.94}$$

obtaining

$$\begin{Bmatrix} dx \\ dy \end{Bmatrix} = j^{-1} \begin{pmatrix} \eta_y & -\xi_y \\ -\eta_x & \xi_x \end{pmatrix} \begin{Bmatrix} d\xi \\ d\eta \end{Bmatrix}, \quad \text{where} \quad j = \xi_x\eta_y - \eta_x\xi_y, \tag{7.95}$$

which implies, if $j \neq 0$, that

$$\begin{pmatrix} \xi_x & \xi_y \\ \eta_x & \eta_y \end{pmatrix} = j \begin{pmatrix} y_\eta & -x_\eta \\ -y_\xi & x_\xi \end{pmatrix}. \tag{7.96}$$

With this proviso, Eqs. (7.91) and (7.92) reduce to the linear, variable-coefficient system

$$y_\eta - x_\eta \tan\theta(\xi, \eta) = 0, \quad y_\xi + x_\xi \cot\theta(\xi, \eta) = 0 \tag{7.97}$$

for the functions $x = \hat{x}(\xi, \eta)$ and $y = \hat{y}(\xi, \eta)$, where $\theta(\xi, \eta)$ is given by (7.93). As explained in the books by Hill and Kachanov, a further transformation reduces these to a pair of linear decoupled telegraph equations, solutions to which are developed in Chapter 6, Section 7 of the former.

The central difficulty with this procedure, of course, is that to determine the stress field in the x, y -plane, the solutions $\hat{x}(\xi, \eta)$ and $\hat{y}(\xi, \eta)$ thus derived must be inverted to obtain the functions $\hat{\xi}(x, y)$ and $\hat{\eta}(x, y)$, the possibility of doing so hinging on the condition $j \neq 0$. Fortunately, the case $j = 0$ excluded by this procedure leads to a number of useful simple solutions, and so we confine ourselves to this case in the remainder of this subsection. Naturally, then, the system (7.97) no longer applies.

To investigate this case we multiply (7.91) by η_y and invoke the condition $j = 0$, together with (7.92), obtaining $\xi_y \eta_y / \sin 2\theta = 0$. Similarly, multiplication of (7.92) by ξ_y and use of (7.91) yields $\xi_x \eta_x / \sin 2\theta = 0$. There are three possibilities: (a) ξ and η are both constant. Then Eqs. (7.89) imply that p and θ are also constant. The slip lines of this *constant state* are orthogonal straight lines. (b) η_x and ξ_y both vanish. Then (7.91) implies that ξ is constant. (c) η_y and ξ_x both vanish, and (7.92) implies that η is constant. In case (c) we use (7.93) to infer that $\xi_x = 2\theta_x$ and $\xi_y = 2\theta_y$, and (7.91) reduces to $\theta_x + (\tan\theta)\theta_y = 0$. With $d\theta = (\theta_x + \theta_y dy/dx)dx$, this, in turn, implies that θ is constant on α-lines. Accordingly, the latter are straight. From (7.89) we have that $p/k = \xi(\beta) - \eta$, so that the stresses are constant along a given α-line ($\beta = const.$), but vary from one such line to another. This field is called a *simple state*. Case (b) is entirely similar and also gives rise to a simple state.

An example of a constant state is afforded by a uniform state of stress in a region adjacent to a straight traction-free boundary. If the boundary has unit normal $\mathbf{i}_1$, say, then T_{11} and T_{12} vanish, and if $T_{22} > 0$, then the larger and smaller in-plane principal stresses are $T_1 = T_{22}$ and $T_2 = 0$. With reference to Problem 7.6 we then have $T_{22} = 2k$ and $p = -k$. The principal stress axes are $\mathbf{u}_1 = \mathbf{i}_2$ and $\mathbf{u}_2 = -\mathbf{i}_1$, and (7.51) may be solved to obtain

$$\mathbf{t} = \tfrac{\sqrt{2}}{2}(\mathbf{i}_1 + \mathbf{i}_2), \quad \mathbf{s} = \tfrac{\sqrt{2}}{2}(\mathbf{i}_2 - \mathbf{i}_1); \quad \text{thus,} \quad \theta = \pi/4. \tag{7.98}$$

Problem 7.11. Use (7.52) to verify that the zero-traction condition, $\mathbf{T}\mathbf{i}_1 = \mathbf{0}$, is satisfied.

To illustrate a simple state, consider the problem of a straight, semi-infinite traction-free crack and the tentative slip-line field sketched in Figure 7.1. This is partitioned into the three regions labeled A, B, and C, in which A and C are constant states and B is a simple state. In region A, T_{22} and T_{12} vanish, and, assuming T_{11} to be positive, $T_{11} = 2k$. We also have $\theta = 3\pi/4$ and $p = -k$.

If we trace a β-line—a curve on which α is constant—from region A into region C, where $\theta = \pi/4$, and make use of $(7.89)_2$, we obtain $-p/2k + \pi/4 = k/2k + 3\pi/4$. Thus, $p = -(1 + \pi)k$ in region C.

Problem 7.12. Use (7.52) to derive the Cartesian stress components $T_{11} = \pi k$, $T_{22} = (\pi + 2)k$, and $T_{12} = 0$ in region C.

The state in region B is similarly obtained by tracing a β-line from region A. Thus, $-p/2k + \varphi = k/2k + 3\pi/4$, where φ is the azimuthal angle in a system of plane polar

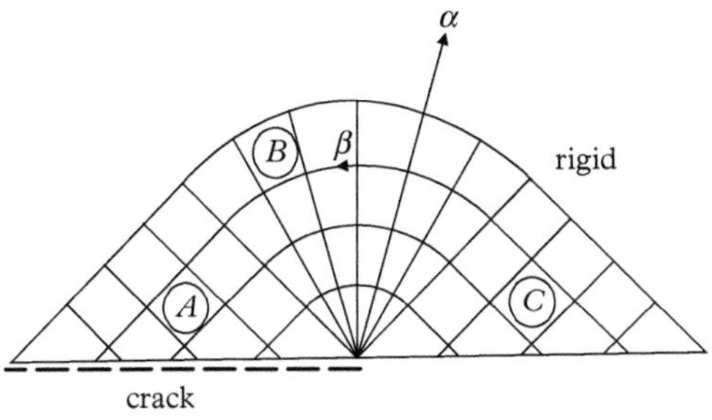

Figure 7.1 *Slip-line field near the tip of a crack. The solution exhibits reflection symmetry with respect to the horizontal dashed line.*

coordinates with origin at the crack tip. Thus,

$$\mathbf{t} = \mathbf{e}_r(\varphi) \quad \text{and} \quad \mathbf{s} = \mathbf{e}_\varphi(\varphi), \tag{7.99}$$

where $\mathbf{e}_r$ and $\mathbf{e}_\varphi$ are defined in Problem 7.5. Thus, in region B we have

$$\theta = \varphi, \quad p = -k(1 + 3\pi/2 - 2\varphi); \quad \pi/4 \leq \varphi \leq 3\pi/4. \tag{7.100}$$

Evidently the stress field in the vicinity of the crack tip is bounded. Note that, as no length scale is specified in this problem, the extent of the plastic regions cannot be determined.

Problem 7.13. Use (7.52) to write the stress in the polar form $\mathbf{T} = T_{rr}\mathbf{e}_r \otimes \mathbf{e}_r + T_{\varphi\varphi}\mathbf{e}_\varphi \otimes \mathbf{e}_\varphi + T_{r\varphi}(\mathbf{e}_r \otimes \mathbf{e}_\varphi + \mathbf{e}_\varphi \otimes \mathbf{e}_r) + T_{zz}\mathbf{k} \otimes \mathbf{k}$. Show that $T_{rr} = T_{\varphi\varphi} = T_{zz} = -p$ and $T_{r\varphi} = k$ in region B.

Problem 7.14. If we reflect this solution through the y-axis passing through the right-most point of the constant state labelled C, we obtain a solution for a body containing two cracks lying on the x-axis with crack tips equidistant from the y-axis and with the slip-line fields meeting at the origin. Assuming the rigidly deforming parts of the body in the region of positive (resp., negative) y to move vertically with velocity V (resp., $-V$), use the Geiringer equations (7.90) to show that a tangential velocity discontinuity occurs on the curves separating the rigid and plastic portions of the body.

7.5.5 Axisymmetric state exterior to a traction-free circular hole

An interesting example of a state that is neither constant nor simple is furnished by the practically important problem of yielded material occupying the region exterior to a traction-free hole. Suppose the hole is circular, with radius a. The exterior unit normal to the boundary is $\mathbf{n} = -\mathbf{e}_r(\varphi)$, and, with the aid of Problem 7.5, the traction condition reduces to

$$T_{rr} = T_{r\varphi} = 0 \quad \text{at} \quad r = a. \tag{7.101}$$

Assuming $T_{\varphi\varphi|r=a}$ to be positive, the ordered principal stresses at $r = a$ are $\{T_1, T_2\} = \{T_{\varphi\varphi}, 0\}$, and therefore, with reference to Problem 7.7, we have $T_{\varphi\varphi|r=a} = 2k$, provided, of course, that the material is in a state of yield at the boundary. Alternatively, $T_{\varphi\varphi|r=a} = -2k$ if $T_{\varphi\varphi|r=a}$ is negative. We consider only the first case, i.e.,

$$T_{\varphi\varphi} - T_{rr} = 2k \quad \text{and} \quad T_{r\varphi} = 0 \quad \text{at} \quad r = a. \tag{7.102}$$

To obtain a solution, we suppose the first of these conditions to hold in the exterior region $r > a$ and assume that T_{rr}, $T_{\varphi\varphi}$, and $T_{r\varphi}$ depend only on r in this region. Because of the simple structure of the assumed solution it is convenient to bypass the Prandtl–Hencky equations and instead appeal directly to the equation of equilibrium, $div\mathbf{T} = \mathbf{0}$, with $div\mathbf{T}$ as given in Problem 7.5. Thus,

$$T'_{r\varphi} + \tfrac{2}{r}T_{r\varphi} = 0 \quad \text{and} \quad T'_{rr} + \tfrac{1}{r}(T_{rr} - T_{\varphi\varphi}) = 0; \quad r > a, \tag{7.103}$$

where $(\cdot)' = d(\cdot)/dr$. The first equation has the solution $T_{r\varphi} = C/r^2$, where C is a constant which vanishes by virtue of the boundary condition. Thus, $T_{r\varphi}(r)$ vanishes for $r \geq a$. Invoking our assumption that $(7.102)_1$ holds for $r > a$, we reduce the second equilibrium equation to $T'_{rr} = 2k/r$, which, together with the boundary condition, gives

$$T_{rr}(r) = 2k\ln(r/a) \quad \text{and} \quad T_{\varphi\varphi}(r) = 2k[1 + \ln(r/a)]; \quad r \geq a. \tag{7.104}$$

This solution meets all conditions stipulated in the problem and furnishes a *post facto* verification of our assumptions. Observe that $T_{rr} > 0$ for all $r>a$. Thus, the material in the region outside a circle of radius r exerts a tensile traction on the material occupying the region $[a, r)$. The alternative case mentioned earlier corresponds to a compressive traction.

To find a velocity field consistent with this solution, we use $p = -\frac{1}{2}(T_{rr} + T_{\varphi\varphi})$ (Why is this true?) to obtain the pressure field

$$p = -k[1 + 2\ln(r/a)]. \tag{7.105}$$

We combine this with our solution, and, making use of $(7.52)_1$, substitute into (7.25) to conclude that the straining tensor is

$$\mathbf{D} = \lambda\boldsymbol{\tau} = \lambda k(\mathbf{e}_\varphi \otimes \mathbf{e}_\varphi - \mathbf{e}_r \otimes \mathbf{e}_r). \tag{7.106}$$

We show that this is compatible with a purely radial velocity field of the form $\mathbf{v} = u(r)\mathbf{e}_r(\varphi)$. Thus,

$$\begin{aligned} \mathbf{L}d\mathbf{y} &= d\mathbf{v} = u'(r)\mathbf{e}_r dr + u\mathbf{e}_r'(\varphi)d\varphi \\ &= u'\mathbf{e}_r dr + u\mathbf{e}_\varphi d\varphi = (u'\mathbf{e}_r \otimes \mathbf{e}_r + \tfrac{u}{r}\mathbf{e}_\varphi \otimes \mathbf{e}_\varphi)d\mathbf{y}, \end{aligned} \tag{7.107}$$

where we have used $dr = \mathbf{e}_r \cdot d\mathbf{y}$ and $rd\varphi = \mathbf{e}_\varphi \cdot d\mathbf{y}$, these following from the formula for $\mathbf{y}$ given in Problem 7.5. The velocity gradient is thus given by the expression in parentheses. Because this is symmetric, it is also the straining tensor, i.e.,

$$\mathbf{D} = u'\mathbf{e}_r \otimes \mathbf{e}_r + \tfrac{u}{r}\mathbf{e}_\varphi \otimes \mathbf{e}_\varphi, \tag{7.108}$$

and comparison with (7.106) furnishes $u/r = \lambda k$ and $u' = -\lambda k$. It follows that u satisfies the differential equation $u' + u/r = 0$, the solution to which is

$$u(r) = C/r, \tag{7.109}$$

with C a constant - positive in accordance with the requirement $\lambda > 0$. The value of this constant is determined by specifying a (positive) value of the velocity at a particular value of the radius. The foregoing stress field is entirely independent of this value, this circumstance being an artifact of the perfect plasticity property $k = const.$

To interpret this solution in term of slip-line theory, we recall that $T_1 = T_{\varphi\varphi}$ and $T_2 = T_{rr}$; then, $\mathbf{u}_1 = \mathbf{e}_\varphi$ and $\mathbf{u}_2 = -\mathbf{e}_r$. Equations (7.51) and (7.54) imply that

$$\theta = \varphi + \pi/4, \tag{7.110}$$

and the equation of an α-line, Eq. $(7.88)_1$, is

$$(\tfrac{dy}{dx})_{|\beta} = \tan(\varphi + \pi/4) = \tfrac{\sin\varphi+\cos\varphi}{\cos\varphi-\sin\varphi} = \tfrac{x+y}{x-y}. \tag{7.111}$$

Recalling that β is constant on such a line, i.e., $\beta_x dx + \beta_y dy = 0$, we have $(\frac{dy}{dx})_{|\beta} = -\beta_x/\beta_y$. Thus, $\beta_x = \beta_y(\frac{x+y}{y-x})$, and the α-lines are such that

$$\tfrac{\beta_y}{y-x}[(x+y)dx + (y-x)dy] = 0. \tag{7.112}$$

The possibility that β_y vanishes must be rejected as it implies that α-lines are the vertical lines on which x assumes constant values; this is inconsistent with (7.110). Thus, the α-lines satisfy the differential equation

$$(x+y)dx + (y-x)dy = 0, \tag{7.113}$$

which, however, is not exact; that is, there is no function β such that $\beta_x = x + y$ and $\beta_y = y - x$. We render it such by introducing a non-zero integrating factor $F(x, y)$, replacing (7.113) with

$$F(x+y)dx + F(y-x)dy = 0, \tag{7.114}$$

and redefining β such that

$$\beta_x = F(x+y) \quad \text{and} \quad \beta_y = F(y-x). \tag{7.115}$$

This may be done without loss of generality; for (7.114) then implies that $d\beta = 0$ on α-lines, which is all that the definition of β requires. The requirement $\beta_{xy} = \beta_{yx}$ then furnishes the partial differential equation

$$(x-y)F_x + (x+y)F_y = -2F, \tag{7.116}$$

the solution to which proceeds by the method of characteristics, whereby $\{x(t), y(t), F(t)\}$ is sought, with

$$\tfrac{dF}{dt} = F_x\tfrac{dx}{dt} + F_y\tfrac{dy}{dt}, \tag{7.117}$$

such that

$$\tfrac{dF}{dt} = -2F, \quad \tfrac{dx}{dt} = x - y \quad \text{and} \quad \tfrac{dy}{dt} = x + y. \tag{7.118}$$

The first of these has the solution $F = F_0 e^{-2t}$, with F_0 a constant, and the second pair combine to give $\frac{d}{dt}(x^2+y^2) = 2(x^2+y^2)$, yielding $2t = \ln r^2 + \mathit{const}$. These furnish the integrating factor $F = C/r^2$, where C is a constant. We observe that this constant cancels out in (7.116), and so it can be specified arbitrarily while fulfilling all restrictions that have been stipulated for the function β. We choose it such that

$$\beta_x = -\tfrac{x+y}{x^2+y^2} \quad \text{and} \quad \beta_y = \tfrac{y-x}{x^2+y^2}. \tag{7.119}$$

To find β we use $r_x = x/r$, which implies that $x/r^2 = (\ln r)_x$; similarly, $y/r^2 = (\ln r)_y$. We also use $\tan\varphi = y/x$ to obtain $(\tan\varphi)_x = -y/x^2$ and $(\tan\varphi)_y = 1/x$; i.e., $\varphi_x \sec^2\varphi = -y/x^2$ and $\varphi_y \sec^2\varphi = 1/x$. These give $\varphi_x = -y/r^2$ and $\varphi_y = x/r^2$, reducing $(7.119)_1$ to $\beta_x = \varphi_x - (\ln r)_x$. Then, $\beta = \varphi - \ln r + f(y)$, where, to satisfy the second of Eqs. (7.119), it is necessary that $f'(y) = 0$. The value of the constant f may be specified by stipulating that the α-line in question intersects the circle $r = a$ at the azimuthal angle φ_0. Because β is constant on this curve we may take $\beta = \varphi_0$, finally reaching

$$\varphi = \beta + \ln(r/a), \tag{7.120}$$

in which $\beta = \varphi_0$ on the considered α-line. The entire family of α-lines is generated by varying the constant β. Proceeding in the same way we find the equations of the β-lines to be

$$\varphi = \alpha - \ln(r/a), \tag{7.121}$$

in which α is the azimuthal angle at which the curve intersects the circle $r = a$. Thus, the slip lines are orthogonal logarithmic spirals. Experimental observations of these spirals are discussed in the books by Nadai and Kachanov.

Problem 7.15. Use the stress field to show that $\pm p/2k + \theta$ are constants on the appropriate logarithmic spirals that define the slip lines.

Problem 7.16. Generalize this problem to accommodate a shear traction $\tau \mathbf{e}_\varphi$ applied at the hole boundary ($r = a$), with $|\tau| < k$. Assuming an axisymmetric stress field, with T_{rr}, $T_{\varphi\varphi}$, and $T_{r\varphi}$ depending only on r, obtain the equilibrium stress distribution assuming the material to be completely yielded. Can you find a velocity field that is consistent with this solution?

Problem 7.17. Investigate the equilibrium state of stress under the assumptions that the stress $T_{r\varphi}$ depends only on φ and the material is in a state of yield.

7.6 Anti-plane shear

This is a much simpler class of problems, exemplified, in the axisymmetric case, by Problem 7.6. More generally, anti-plane shear is associated with velocity fields of the form

$$\mathbf{v} = w(y_1, y_2)\mathbf{k}, \tag{7.122}$$

with gradient

$$\mathbf{L} = \mathbf{k} \otimes grad w, \quad \text{where} \quad grad w = w_{,\alpha}\mathbf{i}_\alpha. \tag{7.123}$$

The straining tensor is

$$\mathbf{D} = \tfrac{1}{2}(\mathbf{k} \otimes grad w + grad w \otimes \mathbf{k}), \tag{7.124}$$

and (7.25) furnishes the deviatoric Cauchy stress

$$\boldsymbol{\tau} = \boldsymbol{\sigma} \otimes \mathbf{k} + \mathbf{k} \otimes \boldsymbol{\sigma}, \quad \text{with} \quad \lambda\boldsymbol{\sigma} = \tfrac{1}{2} grad w. \tag{7.125}$$

The yield condition $|\boldsymbol{\tau}| = \sqrt{2}k$ is equivalent to

$$|\boldsymbol{\sigma}| = k. \tag{7.126}$$

Then, $\lambda = \frac{1}{2k}\,|grad w|$ and

$$\boldsymbol{\sigma} = k\mathbf{t}, \quad \text{where} \quad \mathbf{t} = |grad w|^{-1}\, grad w, \tag{7.127}$$

which is meaningful if the motion is non-rigid, i.e., if $grad w \neq \mathbf{0}$. In this case we introduce a two-dimensional scalar field θ, as in $(7.54)_1$, such that

$$\mathbf{t} = \cos\theta \mathbf{i}_1 + \sin\theta \mathbf{i}_2, \tag{7.128}$$

and use (7.26) to reduce the stress to

$$\mathbf{T} = -p\mathbf{i} + k(\mathbf{t} \otimes \mathbf{k} + \mathbf{k} \otimes \mathbf{t}). \tag{7.129}$$

Assuming equilibrium without body force, and noting that $(grad\mathbf{t})\mathbf{k} = \mathbf{0}$, we arrive at the equation to be solved:

$$grad(p/k) = (div\mathbf{t})\mathbf{k}. \tag{7.130}$$

Problem 7.18. Fill in the steps leading to (7.130).

This yields $p_{,\alpha} = 0$ and $p = P(y_3)$, with $P'(y_3)/k = div\mathbf{t}$. Then, $P''(y_3)$ vanishes and (7.130) reduces to

$$C = k div\mathbf{t}, \tag{7.131}$$

where C is the constant axial pressure gradient in the y_3-direction. This may be cast as an equation for θ. Thus,

$$(grad\mathbf{t})d\mathbf{y} = d\mathbf{t} = \mathbf{s}d\theta = \mathbf{s}(grad\theta \cdot d\mathbf{y}) = (\mathbf{s} \otimes grad\theta)d\mathbf{y}, \quad \text{where} \quad \mathbf{s} = -\sin\theta \mathbf{i}_1 + \cos\theta \mathbf{i}_2, \tag{7.132}$$

implying that $grad\mathbf{t} = \mathbf{s} \otimes grad\theta$ and hence that $div\mathbf{t} = \mathbf{s} \cdot grad\theta$.

Adopting the terminology of plane-strain slip-line theory, we conclude, in the case of a vanishing axial pressure gradient, that $d\theta = 0$ on β-lines, and hence that these lines are straight. Thus,

$$\theta = \eta(\alpha) \tag{7.133}$$

for some function η. Moreover, from $(7.127)_2$ and (7.128) we have that $\mathbf{s} \cdot grad w = 0$. It follows that

$$w = \omega(\alpha), \tag{7.134}$$

for some function ω, and hence that the velocity is constant on a given β-line. Naturally, this constant varies from one such line to another.

To illustrate this model we again consider the field exterior to a traction-free circular hole of radius a. Using (7.129), the zero-traction condition is easily seen to require that

$$p = 0 \quad \text{and} \quad \mathbf{e}_r(\varphi) \cdot \mathbf{t} = 0 \quad \text{at} \quad r = a, \tag{7.135}$$

the first of these implying that $P(y_3) = 0$ and thus that p vanishes indentically in the material region; and the second yielding $\mathbf{t} = \pm \mathbf{e}_\varphi(\varphi)$ and $\mathbf{s} = \mp \mathbf{e}_r(\varphi)$ at $r = a$. We consider the first pair of alternatives for the sake of definiteness.

In fact we obtain a solution simply by extending these conditions into the region $r > a$, i.e., by taking $\mathbf{t} = \mathbf{e}_\varphi(\varphi)$ and $\mathbf{s} = -\mathbf{e}_r(\varphi)$ in the material region. This follows easily from

$$d\mathbf{t} = d\mathbf{e}_\varphi = -\tfrac{1}{r}\mathbf{e}_r(\mathbf{e}_\varphi \cdot d\mathbf{y}), \tag{7.136}$$

yielding $grad\mathbf{t} = -\frac{1}{r}\mathbf{e}_r \otimes \mathbf{e}_\varphi$ and $div\mathbf{t} = -\frac{1}{r}\mathbf{e}_r \cdot \mathbf{e}_\varphi = 0$, as required. The corresponding stress field is given simply by

$$\mathbf{T} = k(\mathbf{e}_\varphi \otimes \mathbf{k} + \mathbf{k} \otimes \mathbf{e}_\varphi), \tag{7.137}$$

and from (7.128) we have $\theta = \varphi + \pi/2$. We may therefore take $\alpha = \varphi$ and $\eta(\alpha) = \alpha + \pi/2$. The condition $\mathbf{s} \cdot grad w = 0$ implies that w is independent of r and hence a function of φ alone, i.e., that $w = \omega(\varphi)$, with

$$grad w = \tfrac{1}{r}\omega'(\varphi)\mathbf{e}_\varphi. \tag{7.138}$$

In the absence of further conditions, this is as far as we can go in solving the problem as stated.

Problem 7.19. Generalize this problem to accommodate a shear traction $\tau\mathbf{k}$ applied at the hole boundary ($r = a$), with $|\tau| < k$. Find an axisymmetric equilibrium stress field in the region $r > a$ and show that it approaches the solution for the traction-free hole as $r/a \to \infty$. What restrictions does the velocity field w satisfy?

This concludes our introduction to the classical theory of the rigid, perfectly plastic response of isotropic materials. We have barely scratched the surface of this large field. The references cited contain many further solutions to a large variety of problems, not limited to conditions of plane or anti-plane strain. Applications to three-dimensional problems, including extensions to account for hardening of the material, may be found in the papers by Durban and Fleck.

References

Batchelor, G. K. (Ed) (1958). *The Scientific Papers of Sir Geoffrey Ingram Taylor*, Vol. 1: *Mechanics of Solids*. Cambridge University Press, Cambridge, UK.

Chadwick, P. (1976). *Continuum Mechanics: Concise Theory and Problems*. Dover, New York.
Christensen, R. M. (2016). *Theory of Materials Failure*. Oxford University Press, Oxford.
Collins, I. F. (1982). Boundary value problems in plane strain plasticity. In H. G. Hopkins and M. J. Sewell (Eds), *Mechanics of Solids: The Rodney Hill 60th Anniversary Volume*, pp. 135–84. Pergamon Press, Oxford.
Cristescu, N. D. (2007). *Dynamic Plasticity*. World Scientific, Singapore.
Durban, D. (1979). Axially symmetric radial flow of rigid/linearly hardening materials. *ASME J. Appl. Mech.* 46, 322–8.
Durban, D. (1980). Drawing of tubes. *ASME J. Appl. Mech.* 47, 736–740.
Durban, D., and Fleck, N. A. (1992). Singular plastic fields in steady penetration of a rigid cone. *ASME J. Appl. Mech.* 59, 706–10.
Edmiston, J., Steigmann, D. J., Johnson, G. J. and Barton, N. (2013). A model for elastic-viscoplastic deformations of crystalline solids based on material symmetry: theory and plane-strain simulations. *Int. J. Eng. Sci.* 63, 10–22.
Epstein, M., and Elżanowski, M. (2007). *Material Inhomogeneities and Their Evolution*. Springer, Berlin.
Geiringer, H. (1973). Ideal plasticity. In C. Truesdell (Ed.) *Mechanics of Solids*, Vol. III, pp. 403–533. Springer, Berlin.
Gupta, A., and Steigmann, D. J. (2014). Plane strain problem in elastically rigid finite plasticity. *Quart. J. Mech. Appl. Math.* 67, 287–310.
Gurtin, M. E. (1981). *An Introduction to Continuum Mechanics*. Academic Press, Orlando, FL.
Hill, R. (1950). *The Mathematical Theory of Plasticity*. Clarendon Press, Oxford.
Johnson, W., and Mellor, P. B. (1980). *Engineering Plasticity*. Van Nostrand Reinhold, London.
Johnson, W., Sowerby, R., and Haddow, J. B. (1970). *Plane-Strain Slip-Line Fields: Theory and Bibliography*. Edward Arnold, London.
Kachanov, L. M. (1974). *Fundamentals of the Theory of Plasticity*. MIR Publishers, Moscow.
Lubliner, J. (2008). *Plasticity Theory*. Dover, New York.
Nadai, A. (1950). *Theory of Flow and Fracture of Solids*. McGraw-Hill, New York.
Neff, P. (2006). A finite-strain elastic-plastic Cosserat theory for polycrystals with grain rotations. *Int. J. Engng. Sci.* 44, 574–94.
Prager, W. (1961). *Introduction to the Mechanics of Continua*. Ginn & Co., Boston.
Truesdell, C., and Noll, W. (1965). The nonlinear field theories of mechanics. In: S. Flügge (Ed.), *Handbuch der Physik III/1*. Springer, Berlin.
Yu, H.-S. (2006). *Plasticity and Geotechnics*. Springer, Berlin.

8

Small-deformation theory

Most of the literature concerned with explicit solutions to problems involving simultaneous elastic–plastic deformations is cast in a framework that presumes small deformations at the outset. The basic theory, due to Prandtl and Reuss, facilitates the appropriation of results from linear elasticity theory, this being by far the most highly developed branch of solid mechanics. To make contact with this substantial literature, it is thus incumbent upon us to develop the small-deformation theory in some detail. Unfortunately, we immediately encounter the difficulty that this theory, as traditionally formulated, fails to be frame invariant, as do traditional formulations of linear elasticity theory. This means that a valid calculation based on the conventional formulation, performed in a frame of reference appropriate to Tokyo, for example, will be invalid in a rotated frame one might use in San Francisco, and that the calculation in Tokyo, if valid at one instant in time, will cease to be valid at other times as the local frame rotates with the Earth. In view of the well-documented success of the small-deformation theories as predictive models, it is thus clear that there must be some conceptual detail in the conventional expositions that is not quite right. The resolution of this difficulty is provided by the *shifter* concept, briefly introduced in Section 2.3. Accordingly we confine ourselves in the present chapter to the development of a frame-invariant formulation of the Prandtl–Reuss theory.

8.1 The displacement field

Students of solid mechanics usually learn about the displacement field prior to the study of more advanced concepts, such as the deformation gradient. It is when encountering the latter in graduate courses that the confusion starts. Typically, the displacement field $\mathbf{u}(\mathbf{x}, t)$ is defined by

$$\mathbf{u}(\mathbf{x}, t) = \mathbf{y} - \mathbf{x}, \quad \text{with} \quad \mathbf{y} = \chi(\mathbf{x}, t), \tag{8.1}$$

in which $\mathbf{y}$ and $\mathbf{x}$ are interpreted, not as positions in point spaces, but rather as position *vectors* in the associated translation, or *tangent* spaces. For example, $\mathbf{y} = y^k\mathbf{i}_k$ in terms of Cartesian coordinates $\{y^k\}$. We then have $\mathbf{y} \in T_{\kappa_t}$, and for consistency it is necessary to interpret $\boldsymbol{\chi}$ as a vector in T_{κ_t}. However, $\mathbf{x} \in T_\kappa$ and (8.1) makes no sense as it stands

A Course on Plasticity Theory. David J. Steigmann, Oxford University Press. © David J. Steigmann (2022).
DOI: 10.1093/oso/9780192883155.003.0008

because it entails the subtraction of vectors belonging to distinct vector spaces. This inconsistency is easily addressed by using the shifter, $\mathbf{1}$, to inject $\mathbf{x}$ into $T_{\mathcal{K}_t}$, the effect of which is to convert $\mathbf{x}$ into an element of $T_{\mathcal{K}_t}$ without otherwise altering it. Its image there is

$$\mathbf{x}^* = \mathbf{1x}, \tag{8.2}$$

and is used to redefine the displacement as

$$\mathbf{u}(\mathbf{x}, t) = \chi(\mathbf{x}, t) - \mathbf{x}^*. \tag{8.3}$$

This definition is meaningful because the right-hand side involves the difference of vectors belonging to the same space.

From (8.2) we have that $\mathbf{x}^*$ is a linear function of $\mathbf{x}$. Its gradient with respect to $\mathbf{x}$ is therefore given by

$$\nabla \mathbf{x}^* = \mathbf{1}. \tag{8.4}$$

Recalling the discussion in Section 2.3 to the effect that $\mathbf{1}$ is a rotation, the fact that it is also a gradient implies that it is spatially uniform. The proof of this assertion proceeds in exactly the same way as the proof of the fact that the rotation accompanying a superposed rigid-body deformation is necessarily uniform. See the proofs in the books by Chadwick and Gurtin, for example. Using convected coordinates $\{\xi^i\}$ to illustrate, we then differentiate (8.2) with respect to the coordinates, obtaining

$$\mathbf{e}_i^* = \mathbf{1e}_i, \tag{8.5}$$

where $\mathbf{e}_i = \mathbf{x}_{,i}$ and $\mathbf{e}_i^* = \mathbf{x}^*_{,i}$. Thus,

$$\mathbf{1} = \mathbf{1I} = \mathbf{1e}_i \otimes \mathbf{e}^i = \mathbf{e}_i^* \otimes \mathbf{e}^i. \tag{8.6}$$

Using $\mathbf{e}^i = \delta^i_j \mathbf{e}^j = \mathbf{e}^j(\mathbf{e}_j^* \cdot \mathbf{e}^{*i}) = (\mathbf{e}^j \otimes \mathbf{e}_j^*)\mathbf{e}^{*i} = \mathbf{1}^t\mathbf{e}^{*i}$, we also have

$$\mathbf{1} = \mathbf{e}_i^* \otimes \mathbf{1}^t\mathbf{e}^{*i} = (\mathbf{e}_i^* \otimes \mathbf{e}^{*i})\mathbf{1} = \mathbf{i1}, \tag{8.7}$$

and thus

$$\mathbf{1I} = \mathbf{i1}. \tag{8.8}$$

From (8.6) and the metric $e_{ij}^* = \mathbf{e}_i^* \cdot \mathbf{e}_j^* = \mathbf{e}_i \cdot \mathbf{e}_j = e_{ij}$, which follows from (8.5) and the fact that $\mathbf{1}$ is a rotation, we also have that

$$\mathbf{1}^t\mathbf{1} = \mathbf{I} \quad \text{and} \quad \mathbf{11}^t = \mathbf{i}, \tag{8.9}$$

both of which have been used on previous occasions, in Sections 2.3 and 7.3, respectively.

Taking the gradient of (8.3), we arrive at

$$\mathbf{F} = \mathbf{1} + \nabla\mathbf{u}, \tag{8.10}$$

where $\mathbf{F} = \nabla\boldsymbol{\chi}$ is the usual deformation gradient. Again this makes sense because each term in this equation is a tensor of the same type; that is, each is a *two-point* tensor that maps T_κ to T_{κ_t}. The traditional expression, $\mathbf{F} = \mathbf{I} + \nabla\mathbf{u}$, is not tensorially meaningful because $\mathbf{I}$ maps T_κ to itself and is therefore not of the same type as $\mathbf{F}$.

The point is reinforced by the example of the superposed rigid-body deformation described in Eq. (2.25). The associated deformation gradient is $\mathbf{QF}$ in which the rotation $\mathbf{Q}$ is spatially uniform. To ensure that (8.10) continues to apply after the rotation we should require that $\mathbf{1}$ be transformed to $\mathbf{Q1}$ and that $\nabla\mathbf{u}$ be transformed to $\mathbf{Q}(\nabla\mathbf{u})$. From (8.5), the first of these is tantamount to the statement that $\mathbf{e}_i^*$ is transformed to $\mathbf{Qe}_i^*$, and hence that the vector space T_{κ_t} is itself rotated by $\mathbf{Q}$, whereas the second implies that $d\mathbf{u} = (\nabla\mathbf{u})d\mathbf{x}$ is transformed to $\mathbf{Q}(\nabla\mathbf{u})d\mathbf{x} = \mathbf{Q}d\mathbf{u} = d(\mathbf{Qu})$, yielding the transformation

$$\mathbf{u} \to \mathbf{Qu} + \mathbf{c}, \tag{8.11}$$

where the vector $\mathbf{c}$ is a function of t. In the same way, from (8.4) we find that $\mathbf{x}^*$ transforms to $\mathbf{Qx}^* + \mathbf{e}$, where the vector $\mathbf{e}$ is a function of t. This is consistent with the conventional notion that $\mathbf{x}$ remains invariant under superposed rigid-body motions provided that $\mathbf{e} = \mathbf{0}$. Taken together with (8.3), these in turn imply that $\boldsymbol{\chi}(\mathbf{x}, t)$ transforms to $\mathbf{Q}(t)\boldsymbol{\chi}(\mathbf{x}, t) + \mathbf{c}(t)$, which is just (2.25).

To interpret (8.11) visually, imagine a displacement vector that is stationary with respect to a fixed frame of reference attached to the ground. Suppose we observe this vector from the vantage point of a passing airplane. From our viewpoint the entire vector undergoes a time-dependent rotation and translation in accordance with (8.11). However, in the traditional interpretation, based on (8.1), we start with (2.25) and, assuming $\mathbf{x}$ to remain invariant, conclude, contrary to the airborne observer, that $\mathbf{u}$ transforms to $\mathbf{Qy} - \mathbf{x} + \mathbf{c}$, and hence that $\nabla\mathbf{u}$ transforms, rather bizarrely, to $\mathbf{Q}(\mathbf{I} + \nabla\mathbf{u}) - \mathbf{I}$. This tensorially meaningless statement lies at the heart of the widespread and persistent misconception that linear elasticity theory is deficient in that it fails to be frame invariant. In contrast, no such difficulty arises when the displacement and its gradient are properly interpreted in terms of (8.3) and (8.10).

This thought experiment suggests that we've been too cavalier about the relationship between frame invariance and invariance under superposed rotations. The former pertains to invariance with respect to a change of observer, whereas the latter pertains to a single observer. In Murdoch's paper it is shown that invariance under superposed rigid-body motions follows from frame invariance. Accordingly the rotation $\mathbf{Q}$ is more properly interpreted as a rotation from one frame of reference to another.

It proves convenient to introduce a purely referential displacement gradient, $D\mathbf{u}$, defined by

$$\nabla\mathbf{u} = \mathbf{1}(D\mathbf{u}), \quad D\mathbf{u} = \mathbf{1}^t(\nabla\mathbf{u}) = \nabla(\mathbf{1}^t\mathbf{u}), \tag{8.12}$$

and to write (8.10) in the form

$$\mathbf{F} = \mathbf{1}(\mathbf{I} + D\mathbf{u}). \tag{8.13}$$

Assuming that the various observers adopt the same reference configuration (so that the operator ∇ remains invariant), from the foregoing discussion we then have that $D\mathbf{u}$ is invariant under a change of frame, this being a trivial consequence of the fact, which follows from (8.11), that $\mathbf{1}^t\mathbf{u} \to \mathbf{1}^t\mathbf{u}+\mathbf{d}(t)$, where $\mathbf{d}(t) = (\mathbf{Q1})^t\mathbf{c}$. Combining this with $(8.9)_1$ and the definition (1.20) of the Lagrange strain $\mathbf{E}$, we derive the strain-displacement relation

$$\mathbf{E} = \boldsymbol{\epsilon} + \tfrac{1}{2}(D\mathbf{u})^t D\mathbf{u}, \quad \text{where} \quad \boldsymbol{\epsilon} = \tfrac{1}{2}\{D\mathbf{u} + (D\mathbf{u})^t\}. \tag{8.14}$$

Note that $\boldsymbol{\epsilon}$ is invariant under a change of frame.

Small-deformation theories are based on the premise that $|D\mathbf{u}| \ll 1$ at all material points. In these circumstances the leading-order approximation $\mathbf{E} \simeq \boldsymbol{\epsilon}$ furnishes a linear strain–displacement relation that is again insensitive to changes of frame; here and henceforth, we use the symbol $\simeq$ to identify expressions that are valid to leading order. In contrast, in conventional treatments the strain–displacement relation is given by (8.14) with $D\mathbf{u}$ replaced by $\nabla\mathbf{u}$. In this case, although $\mathbf{E}$ is invariant under a change of frame, $\boldsymbol{\epsilon}$ is not, and the linear approximation $\mathbf{E} \simeq \boldsymbol{\epsilon}$ yields a strain measure that fails to be invariant. This is the reason why linear elasticity theory, as traditionally formulated, is not frame invariant. Of course no such difficulty arises if proper account is taken of the true nature of the displacement gradient as defined by (8.10).

Naturally the foregoing discussion pertains entirely to kinematics and therefore applies whether or not the material is elastic. However, see the paper by Steigmann for a fuller discussion of this issue as it relates to linear elasticity theory, and the book by Marsden and Hughes for a more general exposition of the shifter concept.

Problem 8.1. Using the convected-coordinate formalism we have $\nabla\mathbf{u} = \mathbf{u}_{,i} \otimes \mathbf{e}^i$ in which $\mathbf{u} \in T_{\mathcal{K}_t}$ in accordance with (8.3). We may thus write $\mathbf{u} = u^j\mathbf{e}^*_j$. Then, $D\mathbf{u} = \mathbf{1}^t\mathbf{u}_{,i} \otimes \mathbf{e}^i = (\mathbf{1}^t\mathbf{u})_{,i} \otimes \mathbf{e}^i$, where $\mathbf{1}^t\mathbf{u} = u^j\mathbf{e}_j$. Show that the components $u^i_{|j}$, where $(\cdot)_{|i}$ is the covariant derivative with respect to the referential connection $\bar{\Gamma}^k_{ij}$, are invariant under a change of frame. Derive the representation (4.12) for the strain, in which

$$E_{ij} = \epsilon_{ij} + \tfrac{1}{2}u^k_{|i}u_{k|j}, \quad \text{with} \quad \epsilon_{ij} = \tfrac{1}{2}(u_{i|j} + u_{j|i}),$$

where $u_i = e_{ij}u^j$ in which $e_{ij} = \mathbf{e}_i \cdot \mathbf{e}_j$ is the referential metric. Show that $\bar{\Gamma}^{*k}_{ij} = \bar{\Gamma}^k_{ij}$.

Problem 8.2. Use (5.2) and (5.3) with (1.20) to obtain

$$\mathbf{E} = \mathbf{G}^t\hat{\mathbf{E}}\mathbf{G} + \hat{\mathbf{P}},$$

where $\hat{\mathbf{E}}$ is the *elastic* strain, given by (5.46), and $\hat{\mathbf{P}}$, the *plastic* strain (not to be confused with the Piola stress), is defined by

$$\hat{\mathbf{P}} = \tfrac{1}{2}(\mathbf{G}^t\mathbf{G} - \mathbf{I}).$$

Show that $\hat{\mathbf{E}} = \hat{E}_{ij}\mathbf{m}^i \otimes \mathbf{m}^j$ and $\hat{\mathbf{P}} = \hat{P}_{ij}\mathbf{e}^i \otimes \mathbf{e}^j$, where $\hat{E}_{ij} = \frac{1}{2}(g_{ij} - m_{ij})$ and $\hat{P}_{ij} = \frac{1}{2}(m_{ij} - e_{ij})$, and thus establish the relation $E_{ij} = \hat{E}_{ij} + \hat{P}_{ij}$, valid for covariant components only.

8.2 Approximations for small displacement gradients

We have already mentioned the leading-order approximation $\mathbf{E} \simeq \boldsymbol{\epsilon}$, valid if $|D\mathbf{u}| \ll 1$, where $\boldsymbol{\epsilon}$ is defined by $(8.14)_2$. This implies that $|\mathbf{E}| \ll 1$, but the converse is not valid. For, if the strain is small, then the polar decomposition of the deformation gradient implies that $\mathbf{F}$ is approximated by a rotation that may differ markedly from $\mathbf{1}$, and this in turn yields a finite displacement gradient. This situation is encountered in thin flexible structures undergoing large deformations while the strains remain small.

Problem 8.3. Use (8.13) and $(8.14)_2$ together with the polar factorization $\mathbf{F} = \mathbf{RU}$, in which $\mathbf{R}$ is a (two-point) rotation tensor and $\mathbf{U}$ is the symmetric, positive definite right-stretch tensor, to show that

$$\mathbf{U} = \mathbf{I} + \boldsymbol{\epsilon} + \mathbf{O}(\varepsilon^2), \quad \mathbf{U}^{-1} = \mathbf{I} - \boldsymbol{\epsilon} + \mathbf{O}(\varepsilon^2) \quad \text{and} \quad \mathbf{R} = \mathbf{1}\{\mathbf{I} + \boldsymbol{\omega} + \mathbf{O}(\varepsilon^2)\},$$

where $\varepsilon = |D\mathbf{u}|$, $|\mathbf{O}(\varepsilon^2)| = O(\varepsilon^2)$ and

$$\boldsymbol{\omega} = \tfrac{1}{2}\{D\mathbf{u} - (D\mathbf{u})^t\}.$$

Accordingly, $\mathbf{U} - \mathbf{I} \simeq \boldsymbol{\epsilon}$ and $\mathbf{R} - \mathbf{1} \simeq \mathbf{1}\boldsymbol{\omega}$ if $\varepsilon \ll 1$. Note that the skew tensor $\boldsymbol{\omega}$, which characterizes the linear-order rotation due to a displacement with a small gradient, is invariant under superposed rigid-body motions.

Problem 8.4. Use (8.13) and (8.14) with $\mathcal{J}_F^2 = \det(\mathbf{F}^t\mathbf{F})$ to obtain

$$\mathcal{J}_F^2 = \det(\mathbf{A} - \lambda\mathbf{I})_{|\lambda=-1}, \quad \text{where} \quad \mathbf{A} = 2\boldsymbol{\epsilon} + (D\mathbf{u})^t D\mathbf{u}.$$

Use the cubic characteristic equation for the eigenvalues of $\mathbf{A}$ to show that

$$\mathcal{J}_F = 1 + tr\boldsymbol{\epsilon} + O(\varepsilon^2),$$

and hence that $\mathcal{J}_F \simeq 1$ if $\varepsilon \ll 1$.

Problem 8.5. Show that

$$\mathbf{F}^* - \mathbf{1} \simeq \mathbf{1}\{(tr\boldsymbol{\varepsilon})\mathbf{I} - (D\mathbf{u})^t\}$$

if $\varepsilon \ll 1$, where $\mathbf{F}^*$, given by (1.36), is the cofactor of the deformation gradient. Thus, $\mathbf{F}^* \simeq \mathbf{1}$.

Problem 8.6. Use (5.3) to conclude, again if $\varepsilon \ll 1$, that $\mathbf{H} \simeq \mathbf{1K}$. Note that this approximation preserves the transformation formulas (6.88) for changes of frame.

Having disposed of the kinematical approximations, we proceed to estimate the relations among the various stress measures and to obtain the approximate equation of motion. For example, if $\varepsilon \ll 1$, which we assume henceforth, then (1.42) and (8.13) imply that

$$\mathbf{P} \simeq \mathbf{1S}, \tag{8.15}$$

where $\mathbf{P}$ and $\mathbf{S}$, respectively, are the Piola and Piola–Kirchhoff stresses. It then follows from (1.38) and the result of Problem 8.5 that the relation between the Cauchy and Piola–Kirchhoff stresses is given, to leading order, by

$$\mathbf{T} \simeq \mathbf{1S1}^t. \tag{8.16}$$

According to (1.27), (1.35), and (1.48), these imply that the Cauchy and Piola tractions, $\mathbf{t}$ and $\mathbf{p}$, respectively, coincide at leading order, i.e.,

$$\mathbf{t} \simeq \mathbf{p} \simeq \mathbf{1}(\mathbf{S}\boldsymbol{\nu}), \tag{8.17}$$

where $\boldsymbol{\nu}$ is the exterior unit normal to the boundary $\partial\pi$ of a material region $\pi \subset \kappa$ on which the traction is acting.

To derive the leading-order approximation to the equation of motion (1.40), we assume, in addition to (8.15), that

$$Div\mathbf{P} \simeq Div(\mathbf{1S}). \tag{8.18}$$

This requires, beyond the small-displacement-gradient assumption $\varepsilon \ll 1$ already made, that the norm of the second gradient of the displacement, $D(D\mathbf{u})$, be small, of the same order as ε.

Problem 8.7. Prove this claim.

These assumptions, however, can only be verified, if true, after a solution to the theory, yet to be formulated, has been obtained. This is exactly the same situation one encounters when using linear elasticity theory. In view of the success of that theory, this issue need hardly deter us from proceeding with the derivation of a parallel formulation of elasto-plasticity theory.

Using convected coordinates, together with the referential counterpart of the formula (3.80) for the divergence, we have $Div\mathbf{P} \simeq Div(\mathbf{1S}) = \{(\mathbf{1S})_{,i}\}\mathbf{e}^i = \mathbf{1}(\mathbf{S}_{,i})\mathbf{e}^i = \mathbf{1}(Div\mathbf{S})$, and the relevant equation of motion, approximating (1.40) at leading order, is

$$Div\mathbf{S} + \rho_{\kappa}\mathbf{1}^t\mathbf{b} \simeq \rho_{\kappa}\mathbf{1}^t\dot{\mathbf{v}}. \tag{8.19}$$

Problem 8.8. A fixed frame of reference used to describe the motion of a body may be regarded as vector space spanned, at a particular point, by $\{\mathbf{e}_i^*\}$ in which the $\mathbf{e}_i^*$ are independent of time. This implies that $\mathbf{1}$ is also independent of time. Use this with $\mathbf{v} = \dot{\boldsymbol{\chi}}$ to show that $\mathbf{1}^t\dot{\mathbf{v}} = \ddot{u}^i\mathbf{e}_i$. In effect, we then regard the vector space T_{κ_t} as being fixed while κ_t itself evolves in the associated point space.

Problem 8.9. It follows easily, from (1.22), (1.32), (1.38), and (1.40), that $J_F div\mathbf{T} = Div(\mathbf{TF}^*)$. Use the divergence and localization theorems with the Piola–Nanson formula (1.35), or direct calculation, to establish the identity

$$J_F div\mathbf{w} = Div\{(\mathbf{F}^*)^t\mathbf{w}\},$$

where $\mathbf{w}$ is a spatial vector field. Show, under our assumptions, that

$$div\mathbf{T} \simeq \mathbf{1}Div(\mathbf{1}^t\mathbf{T1}) \quad \text{and} \quad div\mathbf{w} \simeq Div(\mathbf{1}^t\mathbf{w}).$$

8.3 The frame-invariant Prandtl–Reuss theory

Entirely independently of any assumption about the displacement gradient, in Chapter 6 we formulated the theory of elastic–plastic response on the premise that the norm of the elastic strain $\hat{\mathbf{E}}$ is small. Applying the triangle inequality to the result of Problem 8.2, we have

$$|\mathbf{E}| \leq |\mathbf{G}^t\hat{\mathbf{E}}\mathbf{G}| + |\hat{\mathbf{P}}|. \tag{8.20}$$

Thus, the restriction to small $|\hat{\mathbf{E}}|$ does not imply that $|\mathbf{E}|$ is small. Indeed, this observation underpins the finite-deformation theory of rigid-plastic materials, developed in the previous chapter, in which $|\hat{\mathbf{E}}| = 0$. Moreover, the assumption of a small displacement gradient, while implying that $|\mathbf{E}|$ is small, imposes no restriction on $|\hat{\mathbf{E}}|$ or $|\hat{\mathbf{P}}|$. Accordingly, one may contemplate a model for small displacement gradients having no restrictions on the elastic and plastic deformations beyond those implied by the result of Problem 8.6. Here, however, we proceed, as in the literature, to assume that the norms of both the elastic and plastic strains are small, on the order of that of the displacement gradient. The assumption about the plastic strain implies, as in Problem 8.3, that the right-stretch factor $\mathbf{U}_G$ in the polar factorization $\mathbf{G} = \mathbf{R}_G\mathbf{U}_G$, in which $\mathbf{R}_G$ is a rotation, is approximated, to linear order in $\hat{\mathbf{P}}$, by $\mathbf{I}+\hat{\mathbf{P}}$. We further suppose that $\mathbf{R}_G$ is approximated by $\mathbf{I}+\hat{\boldsymbol{\Omega}}$, where $|\hat{\boldsymbol{\Omega}}| \ll 1$. It then follows that $\hat{\boldsymbol{\Omega}}$ is skew to linear order, and that $\mathbf{G}$ is approximated, again to linear order, by $\mathbf{I}+\hat{\mathbf{P}}+\hat{\boldsymbol{\Omega}}$. Taking this approximation to be exact amounts to assuming that $\dot{\mathbf{G}} = (\hat{\mathbf{P}})^{\cdot} + (\hat{\boldsymbol{\Omega}})^{\cdot}$. We then have that $\dot{\mathbf{G}}\mathbf{G}^{-1}$ is approximated by $(\hat{\mathbf{P}})^{\cdot} + (\hat{\boldsymbol{\Omega}})^{\cdot}$, and the flow rule (6.105) is approximated to consistent order by

$$(\hat{\mathbf{P}})^{\cdot} = \lambda F_{\hat{\mathbf{S}}} + \boldsymbol{\Omega}, \quad \text{with} \quad \boldsymbol{\Omega} = -(\hat{\boldsymbol{\Omega}})^{\cdot}. \tag{8.21}$$

These assumptions are also seen to imply that $\mathbf{G}^t\hat{\mathbf{E}}\mathbf{G}$ is approximated by $\hat{\mathbf{E}}$ and hence, with reference to Problem 8.2, that $\mathbf{E}$ is approximated by $\hat{\mathbf{E}}+\hat{\mathbf{P}}$. With the added assumption of small displacement gradients, we then have

$$\boldsymbol{\epsilon} \simeq \hat{\mathbf{E}} + \hat{\mathbf{P}}, \quad \text{with} \quad \hat{\mathbf{E}} = \mathcal{L}[\hat{\mathbf{S}}], \tag{8.22}$$

where $\mathcal{L}$ is the compliance tensor.

Problem 8.10. With reference to Problem 6.3 and the Appendix to Section 6.1, show that, in the case of isotropy,

$$\mathcal{L}[\hat{\mathbf{S}}] = \tfrac{1}{9\kappa}(tr\hat{\mathbf{S}})\mathbf{I} + \tfrac{1}{2\mu}Dev\hat{\mathbf{S}},$$

where μ and $\kappa(=\lambda+\frac{2}{3}\mu)$ are the (positive) shear and bulk moduli, respectively.

Problem 8.11. Use (6.9) and (6.11) to show, granted our assumptions, that $\hat{\mathbf{S}} \simeq \mathbf{S}$.

Accordingly, assuming, on the basis of (8.22), that $\dot{\boldsymbol{\epsilon}} \simeq (\hat{\mathbf{E}})^{\cdot} + (\hat{\mathbf{P}})^{\cdot}$, and that $(\hat{\mathbf{S}})^{\cdot} \simeq \dot{\mathbf{S}}$, the relevant system is given by (8.17), (8.19) together with

$$Sym(D\mathbf{u})^{\cdot} = \left\{ \begin{matrix} \mathcal{L}[\dot{\mathbf{S}}] + \lambda F_{\mathbf{S}} + \boldsymbol{\Omega}; & \lambda \geq 0, \ \text{ if } \ F(\mathbf{S}) = 0, \\ \mathcal{L}[\dot{\mathbf{S}}]; \ \text{ if } \ F(\mathbf{S}) < 0 & \end{matrix} \right\}. \tag{8.23}$$

This is the frame-invariant extension of the classical Prandtl–Reuss equations, originally stated for isotropic materials, to accommodate anisotropic response. Recalling that the plastic spin may be suppressed in the case of isotropy, the system reduces in this case to

$$Sym(D\mathbf{u})^{\cdot} = \left\{ \begin{matrix} \frac{1}{9\kappa}(tr\dot{\mathbf{S}})\mathbf{I} + \frac{1}{2\mu}Dev\dot{\mathbf{S}} + \lambda Dev\mathbf{S}; & \lambda \geq 0, \ \text{ if } \ |Dev\mathbf{S}| = \sqrt{2}k, \\ \frac{1}{9\kappa}(tr\dot{\mathbf{S}})\mathbf{I} + \frac{1}{2\mu}Dev\dot{\mathbf{S}}; \ \text{ if } \ |Dev\mathbf{S}| < \sqrt{2}k & \end{matrix} \right\}. \tag{8.24}$$

The literature on the Prandtl–Reuss theory for isotropic materials is vast. Many explicit solutions are known, a large number of which are worked out in the books by Hill, Kachanov, and Lubliner, for example, cited in Chapter 7. The book by

Unger documents a number of explicit solutions having immediate relevance to fracture mechanics, and a modern account of the basic theory may be found in Chapter 8 of the book by Bigoni.

Problem 8.12. Use convected coordinates, as in Problem 8.1, to show that $(D\mathbf{u})^{\cdot} = \dot{u}^i_{|j}\mathbf{e}_i \otimes \mathbf{e}^j$.

References

Bigoni, D. (2012). *Nonlinear Solid Mechanics: Bifurcation Theory and Material Instability.* Cambridge University Press, Cambridge, UK.

Chadwick, P. (1976). *Continuum Mechanics: Concise Theory and Problems.* Dover, New York.

Gurtin, M. E. (1981). *An Introduction to Continuum Mechanics.* Academic Press, Orlando, Fl.

Marsden, J. E., and Hughes, T. J. R. (1994). *Mathematical Foundations of Elasticity.* Dover, New York.

Murdoch. A. I. (2003). Objectivity in classical continuum physics: A rationale for discarding the "principle of invariance under superposed rigid-body motions" in favor of purely objective considerations. *Continuum Mech. Thermodyn.* 15, 309–20.

Steigmann, D. J. (2007). On the frame invariance of linear elasticity theory. *ZAMP* 58, 121–36.

Unger, D. J. (1995). *Analytical Fracture Mechanics.* Dover, New York.

9

Strain hardening, rate sensitivity, and gradient plasticity

In this final chapter we extend the theory developed thus far to accommodate strain hardening and also extrapolate Bingham's model to more general models of viscoplasticity. The modeling of strain hardening, in particular, is arguably the main open problem in the phenomenological theory of plasticity. We survey some classical models of hardening and, following this, discuss more recent efforts aimed at modeling observations on the correlation of hardening with sample size, these indicating that material strength is enhanced in sufficiently small metallic bodies relative to larger bodies consisting of the same material. Such scale effects are well documented in the papers by Hutchinson, Stölken and Evans, and Fleck et al. and have given rise to a substantial literature on *gradient plasticity*.

9.1 "Isotropic" hardening

The simplest model of strain hardening is so-called "isotropic" hardening in which the yield stress - the parameter k in von Mises' yield function, for example—is not a constant but instead varies with the state of the material. We say "so-called" because this model is not confined to isotropic material symmetry. Rather than describing a property of isotropic materials, the term instead refers to an overall expansion of the yield surface $F = 0$ that does not alter its shape.

The idea is most easily introduced via von Mises' function (7.10), generalized to

$$F(\hat{\mathbf{S}}, \sigma) = \tfrac{1}{2}\left|Dev\hat{\mathbf{S}}\right|^2 - k^2(\sigma), \tag{9.1}$$

where $k(\sigma)$ is the evolving yield stress in shear and σ is a new scalar variable that evolves with plastic deformation. The models most often considered are associated with one of two evolution equations for σ. These are

$$\dot{\sigma} = \left|\dot{\mathbf{G}}\mathbf{G}^{-1}\right| \quad \text{or} \quad \dot{\sigma} = \hat{\mathbf{S}} \cdot \dot{\mathbf{G}}\mathbf{G}^{-1}, \tag{9.2}$$

A Course on Plasticity Theory. David J. Steigmann, Oxford University Press. © David J. Steigmann (2022).
DOI: 10.1093/oso/9780192883155.003.0009

together with the initial condition $\sigma(\mathbf{x}, t_0) = \sigma_0(\mathbf{x})$. Both models ensure that $\dot{\sigma} > 0$ whenever $\dot{\mathbf{G}} \neq \mathbf{0}$.

Problem 9.1. Verify this claim for $(9.2)_2$. Hint: Use (6.74) and (6.76) together with (6.68) to confirm that $\hat{\mathbf{S}} \cdot \dot{\mathbf{G}}\mathbf{G}^{-1} > 0$, to leading order in the small elastic strain, whenever $\dot{\mathbf{G}} \neq \mathbf{0}$.

Problem 9.2. Show that both models are invariant under $\mathbf{G} \to \mathbf{R}^t\mathbf{G}$ and $\hat{\mathbf{S}} \to \mathbf{R}^t\hat{\mathbf{S}}\mathbf{R}$ for any fixed $\mathbf{R} \in g_{\kappa_i(p)} \subset Orth^+$. Thus, they are meaningful for any type of material symmetry.

Note that both models are also invariant under the time rescaling $t \to ct$ for any positive constant c. They are therefore *rate-independent*. Further, in both cases $\dot{\sigma}$ is independent of the reference configuration κ and invariant under superposed rigid-body motions and time translations $t \to t + a$, for any constant a. Thus, both models are intrinsic to $\kappa_i(p)$ and also frame-invariant. The first alternative is the one most often used in practice. Accordingly, we will confine attention to $(9.2)_1$ henceforth. See Section 2.3 of Hill's book for a fuller discussion of these models, including techniques for the experimental determination of the function $k(\sigma)$.

The variable $\sigma(\mathbf{x}, t)$ reflects the history of plastic flow, and the function $k(\sigma)$ reflects the influence of this history on the current value of the yield stress. Empirical observations typically indicate that $k'(\sigma) > 0$, such that k, like σ, increases with plastic deformation. Hence, the term "strain hardening". Further, this generalization has no effect on the discussion leading to inequality (6.97), apart from the adjustment that F now depends on the current value of σ.

Problem 9.3. Prove this.

Accordingly, the flow rule (6.103) remains in effect with this adjustment, with $F_{\hat{\mathbf{S}}}$ now interpreted as a partial derivative, i.e., as the derivative at fixed σ. This, of course, specializes to (7.14) in the case when F is given by (9.1). Recall that in this flow rule the scalar multiplier field $\lambda(\mathbf{x}, t)$ must be non-negative in accordance with the maximum dissipation inequality (6.97). In the case of perfect plasticity this multiplier is to be regarded as an additional field to be determined in the course of solving the problem at hand. In the case of isotropic hardening, however, we can relate λ to the notions of loading, neutral loading, and unloading as follows.

Thus, suppose there exists an interval of time during which plastic deformation evolves, this being necessary for the existence of a continuous derivative $\dot{\mathbf{G}} \neq \mathbf{0}$. Then $F(\hat{\mathbf{S}}, \sigma)$ vanishes identically in this interval at the material point in question, and hence so too its material derivative:

$$F_{\hat{\mathbf{S}}} \cdot (\hat{\mathbf{S}})^{\cdot} + F_{\sigma}\dot{\sigma} = 0. \tag{9.3}$$

This is called the *consistency condition* for continuing plastic flow. In the case of (9.1) it reduces to

$$Dev\hat{\mathbf{S}} \cdot (\hat{\mathbf{S}})^{\cdot} - 2kk'(\sigma)\dot{\sigma} = 0, \tag{9.4}$$

which specializes, with $(9.2)_1$, to

$$0 = Dev\hat{\mathbf{S}} \cdot (Dev\hat{\mathbf{S}})^{\cdot} - 2kk'(\sigma)\left|\dot{\mathbf{G}}\mathbf{G}^{-1}\right| = Dev\hat{\mathbf{S}} \cdot (Dev\hat{\mathbf{S}})^{\cdot} - 2\lambda kk'(\sigma)\left|Dev\hat{\mathbf{S}}\right|, \tag{9.5}$$

where we have used $\lambda \geq 0$. Solving for λ, we obtain

$$\lambda = \frac{1}{2kk'(\sigma)\left|Dev\hat{\mathbf{S}}\right|} Dev\hat{\mathbf{S}} \cdot (Dev\hat{\mathbf{S}})^{\cdot}, \tag{9.6}$$

which is meaningful, assuming $k'(\sigma) > 0$, provided that

$$\left|Dev\hat{\mathbf{S}}\right| = \sqrt{2}k \quad \text{and} \quad Dev\hat{\mathbf{S}} \cdot (Dev\hat{\mathbf{S}})^{\cdot} \geq 0, \tag{9.7}$$

the latter corresponding, in the case of strict inequality, to *plastic loading*, and, in the case of equality, to *neutral loading*. The remaining possibility, $Dev\hat{\mathbf{S}} \cdot (Dev\hat{\mathbf{S}})^{\cdot} < 0$, is referred to as *unloading*. In this case we stipulate that $\lambda = 0$ and hence that $\dot{\mathbf{G}}$ vanishes. Our terminology refers to a state of yield existing at the onset of the particular loading or unloading condition in question. Thus in the case of loading we have that $(Dev\hat{\mathbf{S}})^{\cdot}$ makes an acute angle with $Dev\hat{\mathbf{S}}$, with $|Dev\hat{\mathbf{S}}| = \sqrt{2}k$. This implies that $\lambda > 0$, $\dot{\mathbf{G}} \neq \mathbf{0}$ and hence $\dot{\sigma} > 0$. With $k'(\sigma) > 0$ it follows that the radius $\sqrt{2}k(\sigma)$ of the hypercylinder defined by $|Dev\hat{\mathbf{S}}| = \sqrt{2}k(\sigma)$ increases monotonically with plastic flow. In the case of neutral loading or unloading we have $\lambda = 0$, $\dot{\mathbf{G}} = \mathbf{0}$, $\dot{\sigma} = 0$, and $k(\sigma)$ remains fixed. This model thus furnishes a faithful representation of the basic phenomenology described in Section 1.1.

The extension to crystalline symmetry is straightforward. Substituting the flow rule (6.105), together with $(9.2)_1$, into the consistency condition (9.3), we obtain

$$0 = F_{\hat{\mathbf{S}}} \cdot (\hat{\mathbf{S}})^{\cdot} + \lambda F_{\sigma}\left|F_{\hat{\mathbf{S}}} + \bar{\mathbf{\Omega}}\right|, \tag{9.8}$$

where $\bar{\mathbf{\Omega}}$ is the skew tensor defined by $\lambda\bar{\mathbf{\Omega}} = \mathbf{\Omega}$. This yields $\mathbf{\Omega} = \mathbf{0}$ when $\dot{\mathbf{G}} = \mathbf{0}$, a condition that requires $\lambda = 0$ by virtue of the orthogonal decomposition (6.105). Moreover, with reference to (6.77) and (6.112), it furnishes a unique $\bar{\mathbf{\Omega}}$ when $\dot{\mathbf{G}} \neq \mathbf{0}$; for we then require that $\lambda > 0$. In other words, we lose no generality if we replace (6.105) by

$$\dot{\mathbf{G}}\mathbf{G}^{-1} = \lambda(F_{\hat{\mathbf{S}}} + \bar{\mathbf{\Omega}}); \quad \lambda \geq 0. \tag{9.9}$$

Further, because the linear space of tensors is the direct sum of the orthogonal subspaces of symmetric and skew tensors, we conclude, from Pythagoras' theorem, that

$$|F_{\hat{\mathbf{S}}} + \bar{\mathbf{\Omega}}| = \sqrt{|F_{\hat{\mathbf{S}}}|^2 + |\bar{\mathbf{\Omega}}|^2}. \tag{9.10}$$

Assuming $F_\sigma < 0$, as in the extended von Mises yield function, we then have

$$\lambda = \frac{1}{|F_\sigma||F_{\hat{\mathbf{S}}}+\bar{\mathbf{\Omega}}|} F_{\hat{\mathbf{S}}} \cdot (\hat{\mathbf{S}})^{\cdot}, \tag{9.11}$$

provided that $F(\hat{\mathbf{S}}, \sigma) = 0$ and $F_{\hat{\mathbf{S}}} \cdot (\hat{\mathbf{S}})^{\cdot} \geq \mathbf{0}$, the latter encompassing plastic loading and neutral loading, whereas $\lambda = 0$ and $\dot{\mathbf{G}} = \mathbf{0}$ in the case $F_{\hat{\mathbf{S}}} \cdot (\hat{\mathbf{S}})^{\cdot} < \mathbf{0}$ associated with unloading.

Isotropic hardening does not account for the Bauschinger effect, described briefly in Section 6.5. Some tentative ideas concerning the modeling of this effect are discussed in Section 9.5, in connection with the theory of *gradient plasticity*.

9.2 Rate sensitivity: Viscoplasticity

Another of our aims is to extend Bingham's model of viscoplasticity to accommodate elastic deformations and crystalline symmetry. The important paper by Perzyna, cited in this chapter's References section, should be consulted for further information on viscoplastic response.

We seek a flow rule in which $\dot{\mathbf{G}}\mathbf{G}^{-1}$ is given explicitly. In the course of adapting Bingham's model (7.29), our first order of business is thus to invert it to obtain an explicit expression for $\mathbf{D}$. To this end we write (7.30) in the form

$$|\mathbf{D}| = \tfrac{1}{2\eta} G(\boldsymbol{\tau}) |\boldsymbol{\tau}|, \quad \text{where} \quad G(\boldsymbol{\tau}) = 1 - \frac{k}{|\boldsymbol{\tau}|/\sqrt{2}}, \tag{9.12}$$

this presuming the value of the yield function $\tilde{F}(\boldsymbol{\tau})$, defined by (7.21), to be non-negative and hence that $G(\boldsymbol{\tau}) \geq 0$. Substituting into (7.29) we then obtain

$$2\eta\mathbf{D} = \left(\frac{G(\boldsymbol{\tau})|\boldsymbol{\tau}|}{G(\boldsymbol{\tau})|\boldsymbol{\tau}| + \sqrt{2}k}\right)\boldsymbol{\tau}; \quad G(\boldsymbol{\tau}) \geq 0, \tag{9.13}$$

which may be simplified on noting that $G(\boldsymbol{\tau})\,|\boldsymbol{\tau}| + \sqrt{2}k = |\boldsymbol{\tau}|$; thus, recalling that $\mathbf{D}$ vanishes if $\tilde{F}(\boldsymbol{\tau}) < 0$, we have

$$2\eta\mathbf{D} = \begin{Bmatrix} G(\boldsymbol{\tau})\boldsymbol{\tau}; & G(\boldsymbol{\tau}) \geq 0, \\ 0; & G(\boldsymbol{\tau}) < 0 \end{Bmatrix}, \quad \text{where} \quad G(\boldsymbol{\tau}) = 1 - \frac{k}{\sqrt{\tilde{F}(\boldsymbol{\tau}) + k^2}}. \tag{9.14}$$

This may be generalized immediately to accommodate elastic–plastic deformations of isotropic materials thus,

$$2\eta\dot{\mathbf{G}}\mathbf{G}^{-1} = \begin{Bmatrix} (1 - \frac{k}{\sqrt{\tilde{F}(Dev\hat{\mathbf{S}})+k^2}})Dev\hat{\mathbf{S}}; & \tilde{F}(Dev\hat{\mathbf{S}}) \geq 0, \\ 0; & \tilde{F}(Dev\hat{\mathbf{S}}) < 0 \end{Bmatrix}, \tag{9.15}$$

where $\tilde{F}(Dev\hat{\mathbf{S}})$ is given by (7.10). See the paper by Atai and Steigmann for an application of this model to the simulation of the transient dynamics of thin metallic sheets.

Extensions of Bingham's model to accommodate crystalline symmetry are much more varied. For one thing, there can be more than one viscosity. Further, most of the models in use are based on the slip-system decomposition (6.20). See the volume edited by Teodosiu, for example. A simple alternative model, merely one among many consistent with the framework discussed in this book, is

$$\dot{\mathbf{G}}\mathbf{G}^{-1} = \begin{Bmatrix} \frac{1}{2\eta}\Phi(F)F_{\hat{\mathbf{S}}} + \mathbf{\Omega}; & F \geq 0, \\ 0; & F < 0 \end{Bmatrix}, \tag{9.16}$$

where η is again a positive viscosity, F is the yield function, and Φ is a positive definite function of F for $F \geq 0$, vanishing at $F = 0$. To ensure continuity at $F = 0$ we write this in the form

$$2\eta\dot{\mathbf{G}}\mathbf{G}^{-1} = \begin{Bmatrix} \Phi(F)(F_{\hat{\mathbf{S}}} + \bar{\mathbf{\Omega}}); & F \geq 0, \\ 0; & F < 0 \end{Bmatrix}, \tag{9.17}$$

where $\mathbf{\Omega} = (\Phi/2\eta)\bar{\mathbf{\Omega}}$. This model is easily seen to be dissipative, as required, provided that $F_{\hat{\mathbf{S}}}$ satisfies inequality (6.113).

The essential feature of all such models is that $\dot{\mathbf{G}}$ is non-zero when $F \geq 0$. In contrast to rate-independent hardening, there is no consistency condition. Nor is there the requirement—a restriction that must be respected in rate-independent plasticity—to ensure that $F \leq 0$ in the course of integrating the flow rule. From the viewpoint of numerical analysis, these are major advantages of viscoplasticity theory over the rate-independent theory.

9.3 Scale effects

A natural way to model length-scale effects in the plastic deformation of metals is to include the gradient $\nabla\mathbf{G}$ of the plastic deformation as an argument of the relevant constitutive functions. This has dimensions of reciprocal length, meaning that a *material* length scale must be present in the non-dimensionalized constitutive functions. Because the constitutive functions pertain to material points, this scale is necessarily representative of the fine structure of the material. Accordingly, $\nabla\mathbf{G}$ exerts influence on the response when the plastic deformation varies significantly over a length scale comparable to the local material scale.

In our work we have discussed only two constitutive functions, the strain-energy function and the yield function, both of which are, of course, scalar-valued. Recall that these have been defined intrinsically, to reflect the material properties existing in $\kappa_i(p)$, and are therefore entirely independent of a reference configuration. This codifies the view that the choice of reference is largely arbitrary, more a matter of convenience than anything else, and so should not influence physical properties as encoded in the undistorted state of the material. While we naturally intend to retain this feature of constitutive functions in the extension to scale-dependent response, we must now contend with the fact that both $\mathbf{G}$ and $\nabla\mathbf{G}$, its gradient with respect to $\mathbf{x} \in \kappa$, are very much dependent on the choice of reference configuration.

We will prove the remarkable fact that any scalar-valued constitutive function, $\mathcal{F}(\mathbf{G}, \nabla\mathbf{G})$ say, that is insensitive to the choice of reference configuration, is necessarily a function of the true dislocation density $\boldsymbol{\alpha}$ defined by (5.20), i.e.,

$$\mathcal{F}(\mathbf{G}, \nabla\mathbf{G}) = f(\boldsymbol{\alpha}), \tag{9.18}$$

for some function f. Sufficiency follows immediately from the fact, proved in Section 5.1, that $\boldsymbol{\alpha}$ is invariant under a change of reference configuration $\kappa \to \bar{\kappa}$ defined by $\bar{\mathbf{x}} = \boldsymbol{\lambda}(\mathbf{x})$ for any smooth function $\boldsymbol{\lambda}$, where $\mathbf{x}$ is the position of a material point p in κ, and $\bar{\mathbf{x}}$ its position in $\bar{\kappa}$ (in Chapter 5 we used the notation $\mathbf{x}_1 = \mathbf{x}$ and $\mathbf{x}_2 = \bar{\mathbf{x}}$). The proof of necessity, which is more difficult, is facilitated by using convected coordinates.

Proceeding, we obtain a useful expression for the third-order tensor $\nabla\mathbf{G}$ on noting that

$$(\nabla\mathbf{G})d\mathbf{x} = d\mathbf{G} = \mathbf{G}_{,i}d\xi^i = \mathbf{G}_{,i}(\mathbf{e}^i \cdot d\mathbf{x}) = (\mathbf{G}_{,i} \otimes \mathbf{e}^i)d\mathbf{x}, \tag{9.19}$$

where we have abused the notation, albeit harmlessly, by writing $\mathbf{G}(\xi^i, t) = \mathbf{G}(\mathbf{x}(\xi^i), t)$, and, of course, the commas are partial derivatives with respect to the coordinates. Thus,

$$\nabla\mathbf{G} = \mathbf{G}_{,i} \otimes \mathbf{e}^i, \quad \text{where} \quad \mathbf{G} = \mathbf{m}_j \otimes \mathbf{e}^j, \tag{9.20}$$

the second of these being simply a restatement of (5.24). We use this with (5.29) and (3.66)—the latter in the form $\mathbf{e}^j_{,i} = -\Gamma^j_{(\kappa)ki}\mathbf{e}^k$ (where $\Gamma^j_{(\kappa)ki}$ are the same as the $\bar{\Gamma}^j_{ki}$ of Chapter 4)—to arrive at

$$\nabla\mathbf{G} = (\mathbf{m}_{j,i} \otimes \mathbf{e}^j + \mathbf{m}_j \otimes \mathbf{e}^j_{,i}) \otimes \mathbf{e}^i = (\hat{\Gamma}^k_{ji} - \Gamma^k_{(\kappa)ji})\mathbf{m}_k \otimes \mathbf{e}^j \otimes \mathbf{e}^i. \tag{9.21}$$

This, of course, pertains to the use of κ as reference. Using $\bar{\kappa}$ instead gives

$$\bar{\mathbf{G}} = \mathbf{m}_j \otimes \bar{\mathbf{e}}^j \quad \text{and} \quad \bar{\nabla}\bar{\mathbf{G}} = (\hat{\Gamma}^k_{ji} - \Gamma^k_{(\bar{\kappa})ji})\mathbf{m}_k \otimes \bar{\mathbf{e}}^j \otimes \bar{\mathbf{e}}^i, \tag{9.22}$$

where $\bar{\mathbf{e}}^i$ are the reciprocals of $\bar{\mathbf{e}}_i = \bar{\mathbf{x}}_{,i}$. The presumed insensitivity of our generic constitutive function to the choice of reference configuration is thus equivalent to the constraint

$$\mathcal{F}(\mathbf{G}, \nabla\mathbf{G}) = \mathcal{F}(\bar{\mathbf{G}}, \bar{\nabla}\bar{\mathbf{G}}). \tag{9.23}$$

It follows, exactly as in the derivation of the representation (4.9) of the deformation gradient, that

$$\nabla\boldsymbol{\lambda} = \bar{\mathbf{e}}_i \otimes \mathbf{e}^i, \tag{9.24}$$

yielding $(\nabla\boldsymbol{\lambda})^{-1} = \mathbf{e}_i \otimes \bar{\mathbf{e}}^i$, $(\nabla\boldsymbol{\lambda})^{-t} = \bar{\mathbf{e}}^i \otimes \mathbf{e}_i$, and, thus,

$$\bar{\mathbf{e}}^j = (\nabla\boldsymbol{\lambda})^{-t}\mathbf{e}^j. \tag{9.25}$$

Further, again with a slight abuse of notation, and assuming that $\boldsymbol{\lambda}$ is twice differentiable with respect to the coordinates (so that the connection $\Gamma^k_{(\bar{\mathcal{K}})ji}$ is well defined), we have

$$\Gamma^k_{(\bar{\mathcal{K}})ji} = \bar{\mathbf{e}}^k \cdot \bar{\mathbf{x}}_{,ji} = (\nabla\boldsymbol{\lambda})^{-t}\mathbf{e}^k \cdot \boldsymbol{\lambda}_{,ji}, \tag{9.26}$$

which, of course, is symmetric in the subscripts.

In Section 6.4 we made essential use of the arbitrariness in the choice of reference, i.e., the arbitrariness of the map $\boldsymbol{\lambda}$. We do so again here, choosing $\boldsymbol{\lambda}(\mathbf{x})$ such that $\nabla\boldsymbol{\lambda} = \mathbf{G}$ *at* the material point p in question, *at* a fixed instant in time; then, $\bar{\mathbf{e}}^k = \mathbf{m}^k$ and $\bar{\mathbf{G}} = \mathbf{I}$ (see (5.13)). In addition, noting that the first and second gradients of any map may be specified independently at any one point, we choose $\boldsymbol{\lambda}(\mathbf{x})$ such that $\Gamma^k_{(\bar{\mathcal{K}})ji}(= \mathbf{m}^k \cdot \boldsymbol{\lambda}_{,ji}) = \hat{\Gamma}^k_{(ji)}$ *at* the point and instant in question, this being possible by virtue of the symmetry of $\Gamma_{(\bar{\mathcal{K}})}$ in the subscripts. (See the paper by Cermelli and Gurtin for an explicit example of such a map.) This results in

$$\bar{\nabla}\bar{\mathbf{G}} = \hat{\mathbf{T}}, \quad \text{where} \quad \hat{\mathbf{T}} = \hat{T}^k_{\cdot ji}\mathbf{m}_k \otimes \mathbf{m}^j \otimes \mathbf{m}^i, \tag{9.27}$$

in which $\hat{T}^k_{\cdot ji} = \hat{\Gamma}^k_{[ji]}$ is the torsion of the Weitzenböck connection $\hat{\Gamma}^k_{ji}$. We thus reduce (9.23) to

$$\mathcal{F}(\mathbf{G}, \nabla\mathbf{G}) = \mathcal{F}(\mathbf{I}, \hat{\mathbf{T}}) = \tilde{\mathcal{F}}(\hat{\mathbf{T}}), \quad \text{say}. \tag{9.28}$$

We have seen, in Section 5.2, that $\hat{\mathbf{T}}$ stands in one-to-one relation to the dislocation density $\boldsymbol{\alpha}$, and (9.18) thus follows from (9.28). Because of the arbitrariness of the material point and time instant in question, this result holds at all such points and at all times.

9.3.1 Crystalline symmetry

Naturally, our constitutive function must conform to requirements imposed by material symmetry. It is thus invariant under $\mathbf{K} \to \mathbf{KR}$; equivalently, $\mathbf{G} \to \mathbf{R}^t\mathbf{G}$, for all $\mathbf{R} \in g_{\mathcal{K}_i(p)} \subset Orth^+$. Recalling the discussion of Section 6.6, in the case of crystalline symmetry the elements of $g_{\mathcal{K}_i(p)}$ are spatially uniform.

Problem 9.4. If $\mathbf{A}$ is uniform and invertible, show that $Curl(\mathbf{A}^{-1}\mathbf{K}^{-1}) = (Curl\mathbf{K}^{-1})\mathbf{A}^{-t}$.

Thus, on replacing $\mathbf{K}$ by $\mathbf{KR}$ in (5.20), we find that

$$\boldsymbol{\alpha} \to \mathbf{R}^t\boldsymbol{\alpha}\mathbf{R}, \tag{9.29}$$

and hence that the constitutive function $f(\boldsymbol{\alpha})$ must be such that

$$f(\boldsymbol{\alpha}) = f(\mathbf{R}^t\boldsymbol{\alpha}\mathbf{R}); \quad \mathbf{R} \in g_{\kappa_i(p)}. \tag{9.30}$$

For example, any function of the form $f(\boldsymbol{\alpha}) = g(|\boldsymbol{\alpha}|)$ automatically meets this requirement, for any orthogonal $\mathbf{R}$.

Another way to arrive at this result is to note, from $(9.22)_2$, that $\mathbf{m}_i \to \mathbf{R}^t\mathbf{m}_i$ and, from $\mathbf{K} \to \mathbf{KR}$ with $\mathbf{K} = \mathbf{e}_i \otimes \mathbf{m}^i$, that $\mathbf{m}^i \to \mathbf{R}^t\mathbf{m}^i$ for all rotations $\mathbf{R}$ belonging to $g_{\kappa_i(p)}$. This implies, for uniform $\mathbf{R}$, that the Weitzenböck connection $\hat{\Gamma}^k_{ij}$ is preserved under symmetry transformations, i.e.

$$\hat{\Gamma}^k_{ij} \to \mathbf{R}^t\mathbf{m}^k \cdot (\mathbf{R}^t\mathbf{m}_i)_{,j} = \mathbf{R}^t\mathbf{m}^k \cdot \mathbf{R}^t\mathbf{m}_{i,j} = \mathbf{m}^k \cdot \mathbf{R}\mathbf{R}^t\mathbf{m}_{i,j} = \mathbf{m}^k \cdot \mathbf{m}_{i,j} = \hat{\Gamma}^k_{ij}. \tag{9.31}$$

It follows that the torsion $\hat{T}^k_{\cdot ij} = \hat{\Gamma}^k_{[ij]}$ is invariant under symmetry transformations, and therefore that it is uniquely defined for any kind of crystalline symmetry once a coordinate system has been selected. The torsion therefore furnishes an invariant measure of defect content, or inhomogeneity, in crystalline solids.

Note that length-scale effects in crystalline materials, represented by functions of the form $\mathcal{F}(\mathbf{G}, \nabla\mathbf{G})$, are operative if and only if the plastic deformation is incompatible, i.e., $Curl\mathbf{G} \neq \mathbf{0}$.

9.3.2 Isotropy

The situation is quite different for isotropic materials. Again with reference to Section 6.6, recall that for these there is no requirement that the rotations $\mathbf{R}$ be uniformly distributed. For example, the scalars $I_{1,2,3}$ comprising the arguments of the strain-energy function for an isotropic solid, defined in the Appendix to Section 6.1, are invariant under any rotation $\mathbf{R}$, uniform or not. For non-uniform $\mathbf{R}$, in place of (9.31) we find that

$$\hat{\Gamma}^k_{ij} \to \hat{\Gamma}^k_{ij} + \mathbf{R}^t\mathbf{m}^k \cdot \mathbf{R}^t_{,j}\mathbf{m}_i. \tag{9.32}$$

The torsion thus transforms as $\hat{T}^k_{\cdot ij} \to \hat{T}^k_{\cdot ij} + \mathbf{R}^t\mathbf{m}^k \cdot \mathbf{R}^t_{,[j}\mathbf{m}_{i]}$, and the components of the dislocation density as $\alpha^{ij} \to \alpha^{ij} + \beta^{ij}$, where $\beta^{ij} = \hat{\varepsilon}^{kli}\mathbf{R}^t\mathbf{m}^j \cdot \mathbf{R}^t_{,[l}\mathbf{m}_{k]}$. In place of (9.30) we have

$$f(\boldsymbol{\alpha}) = f(\mathbf{R}^t(\boldsymbol{\alpha} + \boldsymbol{\beta})\mathbf{R}), \tag{9.33}$$

for any rotation $\mathbf{R}$, which, for $\mathbf{R} = \mathbf{I}$, yields $f(\boldsymbol{\alpha}) = f(\boldsymbol{\alpha} + \boldsymbol{\beta})$, and hence—because $\boldsymbol{\beta}$ can assume arbitrary values—the conclusion that f is necessarily independent of the dislocation density. Of course a dislocation density can be computed, but it cannot enter the constitutive functions as an independent variable in the case of isotropy. This conclusion is natural in view of the fact that dislocations are associated intimately with crystal lattices, whereas isotropic materials are not crystalline. It follows from this line of reasoning that constitutive functions of the form $\mathcal{F}(\mathbf{G}, \nabla\mathbf{G})$ do not characterize isotropic materials.

To explore this issue further, we recall that because of material symmetry, the field $\mathbf{K}$ is non-unique to the extent that no distinction can be made between $\mathbf{K}$ and $\mathbf{KR}$ as far as mechanical response is concerned, where, in the case of isotropy, $\mathbf{R}(\mathbf{x}, t)$ is any rotation field whatsoever. Equivalently, no distinction can be made between $\mathbf{G}$ and $\mathbf{R}^t\mathbf{G}$. We can therefore select $\mathbf{R}$ to be the rotation $\mathbf{R}_G$ in the polar factorization $\mathbf{G} = \mathbf{R}_G\mathbf{U}_G$, in which $\mathbf{U}_G$ is the symmetric, positive definite plastic right-stretch tensor, to conclude that $\mathbf{G}$ and $\mathbf{U}_G$ are mechanically indistinguishable. Further, $\mathbf{U}_G$ is the unique root of $\mathbf{U}_G^2 = \mathbf{M}$, where

$$\mathbf{M} = \mathbf{G}^t\mathbf{G} = m_{ij}\mathbf{e}^i \otimes \mathbf{e}^j \tag{9.34}$$

is the plastic Cauchy–Green tensor (see (5.25)). Conversely, if $\mathbf{G}$ and $\mathbf{R}^t\mathbf{G}$ yield the same value of $\mathbf{M}$, then $\mathbf{R}$ is an arbitrary rotation (see Problem 2.1) and hence an element of the material symmetry group in the case of isotropy. Accordingly, for isotropic materials $\mathbf{G}$ and $\mathbf{M}$ are equivalent descriptors of plastic deformation.

Problem 9.5. Use (9.34) to determine $\dot{\mathbf{M}}$ in terms of $\dot{\mathbf{G}}$. We have shown that $\mathbf{G}$ and $\mathbf{M}$ are mechanically equivalent in isotropic materials. It should therefore be possible to determine $\dot{\mathbf{G}}$ in terms of $\dot{\mathbf{M}}$. Show that this is indeed the case if we take $\dot{\mathbf{G}}\mathbf{G}^{-1}$ to be symmetric, as stipulated, for isotropic materials, in Chapter 7. Write the flow rule (7.8) in the form $\dot{\mathbf{M}} = \dots$.

In view of this observation, a natural course to follow when contemplating scale-dependent constitutive functions for isotropic materials would be to adapt the development for crystalline materials by considering constitutive functions of the form $\mathcal{F}(\mathbf{M}, \nabla\mathbf{M})$. However, as any function of $\mathbf{M}$ and $\nabla\mathbf{M}$ is expressible as a function of $\mathbf{G}$ and $\nabla\mathbf{G}$, and thus ultimately—if also insensitive to the reference configuration—as a function of $\boldsymbol{\alpha}$, this option must be dismissed for the reasons already given. Another way to arrive at this conclusion is to note that

$$\nabla\mathbf{M} = m_{ij|k}\mathbf{e}^i \otimes \mathbf{e}^j \otimes \mathbf{e}^k \tag{9.35}$$

may be reduced to the (third-order) zero tensor by a map $\boldsymbol{\lambda}$ from κ to $\bar{\kappa}$, constructed such that $\Gamma^k_{(\bar{\kappa})ij} = \left\{ {k \atop ij} \right\}$ at the material point in question, where $\left\{ {k \atop ij} \right\}$ is the Levi-Civita connection based on the metric m_{ij} (see (5.35)). This is again permissible because $\Gamma_{(\bar{\kappa})}$ is symmetric in the subscripts. The claim then follows from the fact that every Levi-Civita connection is compatible with its own metric, so that Ricci's lemma is then operative. The gradient $\bar{\nabla}\bar{\mathbf{M}}$, computed using $\bar{\kappa}$ as reference, is thereby reduced to zero at the point in question. Combining this with the fact that $\bar{\mathbf{M}}$ may be reduced to $\mathbf{I}$ at the same point, the presumed invariance of the constitutive function under an arbitrary change of reference then yields $\mathcal{F}(\mathbf{M}, \nabla\mathbf{M}) = \mathcal{F}(\mathbf{I}, \mathbf{0})$, and the conclusion follows.

The obvious next step in the search for a model of scale-dependent plasticity in isotropic materials is to consider constitutive functions that depend on $\mathbf{M}$, $\nabla\mathbf{M}$, and $\nabla\nabla\mathbf{M}$. However, rather than pursue this option in full generality, we follow the example of the model for crystalline materials in which length-scale effects are due entirely to the incompatibility of the plastic deformation. Extending this notion to isotropic materials, we thus seek a model that is sensitive to the incompatibility of $\mathbf{M}$, or, equivalently, to that of the plastic strain $\frac{1}{2}(\mathbf{M} - \mathbf{I})$. We have encountered essentially the same issue in Section 4.2, concluding there that strain compatibility is associated with the vanishing of the relevant Riemann tensor. The appropriate descriptor of *in*compatibility in the present instance is therefore the (generally non-vanishing) Riemann tensor constructed from the Levi-Civita connection $\{^{\,k}_{ij}\}$, with components

$$\mathcal{R}^{k}_{.mlj} = \{^{\,k}_{mj}\}_{,l} - \{^{\,k}_{ml}\}_{,j} + \{^{\,k}_{il}\}\{^{\,i}_{mj}\} - \{^{\,k}_{ij}\}\{^{\,i}_{ml}\}. \tag{9.36}$$

This involves the metric and its partial coordinate derivatives through the second order, and accordingly accounts for plastic length-scale effects. The geometry of the material manifold is thus Riemannian in the case of isotropic materials, in contrast to the Weitzenböckian geometry of crystalline materials.

Problem 9.6. The connection $\{^{\,k}_{ij}\}$ also incorporates length-scale effects because it involves the metric and its first-order coordinate derivatives. Would it be reasonable to propose a constitutive function having $\{^{\,k}_{ij}\}\mathbf{m}_k \otimes \mathbf{m}^i \otimes \mathbf{m}^j$ as its only argument?

Problem 9.7. What is the form of the function $\mathcal{F}(\mathbf{M}, \nabla\mathbf{M}, \nabla\nabla\mathbf{M})$ which is such as remain invariant under an arbitrary thrice-differentiable map $\bar{\mathbf{x}} = \boldsymbol{\lambda}(\mathbf{x})$?

We have seen, in Section 4.2, that the Riemann tensor associated with any Levi-Civita connection possesses certain symmetries that, in three dimensions, render it equivalent to the associated (symmetric) Ricci tensor. Accordingly, any function of the components (9.36) is a (different) function of the Ricci tensor components (see (4.36))

$$\rho_{ij} = \mathcal{R}^{k}_{.jik}. \tag{9.37}$$

Using these we construct the symmetric tensor

$$\boldsymbol{\rho} = \rho_{ij}\mathbf{m}^i \otimes \mathbf{m}^j, \tag{9.38}$$

which is manifestly independent of any reference configuration. Accordingly, a constitutive function of the form

$$\mathcal{F} = f(\boldsymbol{\rho}) \tag{9.39}$$

is automatically insensitive to the choice of reference configuration.

To explore the extent to which such a function is consistent with material symmetry, we note that the metric $m_{ij}(= \mathbf{m}_i \cdot \mathbf{m}_j)$ remains *globally* invariant under symmetry transformations; that is, it is invariant under $\mathbf{m}_i \to \mathbf{R}^t\mathbf{m}_i$ with $\mathbf{R}(\mathbf{x}, t)$ an arbitrary rotation field, for all $\mathbf{x} \in \kappa$. This means that its partial derivatives with respect to the coordinates ξ^i are likewise invariant, and hence so too the connection $\{^{k}_{ij}\}$ and its derivatives. Thus, the components ρ_{ij} are invariant; then, because $\mathbf{m}^i \to \mathbf{R}^t\mathbf{m}^i$, material symmetry considerations require that

$$f(\boldsymbol{\rho}) = f(\mathbf{R}^t\boldsymbol{\rho}\mathbf{R}) \quad \text{for all rotations} \quad \mathbf{R}. \tag{9.40}$$

9.4 Scale-dependent yielding

The simplest way to incorporate scale effects into plasticity theory is to modify the yield function accordingly. Thus, for crystalline materials we would propose that $F(\hat{\mathbf{S}})$ be replaced by a function $F(\hat{\mathbf{S}}, \boldsymbol{\alpha})$, subject to

$$F(\hat{\mathbf{S}}, \boldsymbol{\alpha}) = F(\mathbf{R}^t\hat{\mathbf{S}}\mathbf{R}, \mathbf{R}^t\boldsymbol{\alpha}\mathbf{R}); \quad \mathbf{R} \in g_{\mathcal{K}_i(p)}. \tag{9.41}$$

For isotropic materials we could replace $F(\hat{\mathbf{S}})$ by a function $F(\hat{\mathbf{S}}, \boldsymbol{\rho})$, subject to

$$F(\hat{\mathbf{S}}, \boldsymbol{\rho}) = F(\mathbf{R}^t\hat{\mathbf{S}}\mathbf{R}, \mathbf{R}^t\boldsymbol{\rho}\mathbf{R}), \quad \text{for all rotations} \quad \mathbf{R}. \tag{9.42}$$

For example, we could consider yield functions of the form $F = H(\hat{\mathbf{S}}) - k^2$, where k is the scale-dependent yield stress, a function of $\boldsymbol{\alpha}$ or $\boldsymbol{\rho}$, as appropriate. Indeed, in the case of crystalline materials, such a proposal may be motivated by G. I. Taylor's work in which the flow stress is found to be related to dislocation content.

We retain the assumption—appropriate for rate-independent response—that plastic evolution, i.e., $\dot{\mathbf{K}} \neq \mathbf{0}$, is possible only when $F = 0$, and further assume the stress $\hat{\mathbf{S}}$ to be always confined to the *current elastic range* defined by $F(\cdot, \boldsymbol{\alpha}) \leq 0$ or $F(\cdot, \boldsymbol{\rho}) \leq 0$, according as the material is crystalline or isotropic, respectively. In view of our restriction to materially uniform bodies, we require, as assumed previously, that F be the same function at all material points. The considerations about cyclic processes pertaining to the work inequality (6.90), culminating in (6.103) and (6.104), are seen to carry over unchanged, provided that the function F is continuous and $\boldsymbol{\alpha}$ (or $\boldsymbol{\rho}$, as appropriate) is a continuous function of time during the cycle. Because $\boldsymbol{\alpha}$ involves *Curl*$\mathbf{G}$, which may be specified independently of $\mathbf{G}$ at a given material point, such continuity is an additional assumption beyond the continuity of $\mathbf{K}$ assumed previously. Exceptionally, the continuity of $\boldsymbol{\alpha}$ follows from that of $\mathbf{K}$ for cycles in which $Curl\mathbf{G}$ remains fixed. Invoking (6.90) for such cycles, we recover (6.103) and (6.104), as before, but with F replaced by $F(\mathbf{S},\boldsymbol{\alpha})$. Accordingly, the flow rule (6.105) remains in effect with this adjustment.

This flow rule generally predicts a non-zero value of $Curl\dot{\mathbf{G}}$ in a given initial-boundary-value problem, and therefore appears, at first glance, to be inconsistent with the assumption $Curl\dot{\mathbf{G}} = \mathbf{0}$ just invoked in the course of deriving it. A little reflection is sufficient to overcome this objection, however. For the work inequality pertains to a material point, whereas the *post facto* evaluation of $Curl\dot{\mathbf{G}}$ from the flow rule, at the *same* point, requires the function $\dot{\mathbf{G}}(\mathbf{x}, t)$ for $\mathbf{x}$ in a *neighborhood* of the point in question. Moreover, the work inequality (6.90) purports to apply to all cyclic processes, and hence to those restricted as indicated. Beyond this, the cyclic process associated with the work inequality need have no relation to the actual process that the flow rule purports to describe.

In the case of isotropy we can similarly arrange a cyclic process in which the variation of $\boldsymbol{\rho}$ is controlled entirely by the variation of $\mathbf{K}$ at a given material point, so that continuity of the latter implies that of the former. Invoking (6.90) for such cycles, we arrive at (6.105), with F replaced by $F(\hat{\mathbf{S}}, \boldsymbol{\rho})$ and with vanishing plastic spin. A simple model of this kind is discussed in the paper by Krishnan and Steigmann.

Problem 9.8. Prove these assertions.

To determine the plastic multiplier λ appearing in the flow rule, we need to invoke the relevant consistency condition. In the case of crystalline materials, for example, this is

$$F_{\hat{\mathbf{S}}} \cdot (\hat{\mathbf{S}})^{\cdot} + F_{\boldsymbol{\alpha}} \cdot \dot{\boldsymbol{\alpha}} = 0, \tag{9.43}$$

where, from (5.20),

$$\dot{\boldsymbol{\alpha}} = J_G^{-1}\mathbf{G}(Curl\dot{\mathbf{G}}) + (\dot{\mathbf{G}}\mathbf{G}^{-1})\boldsymbol{\alpha} - (\dot{J}_G/J_G)\boldsymbol{\alpha}, \quad \text{with} \quad \dot{J}_G/J_G = tr(\dot{\mathbf{G}}\mathbf{G}^{-1}). \tag{9.44}$$

Substitution of (6.105) in (9.43) results in a linear constraint on λ and $\nabla\lambda$ jointly, furnishing a first-order partial differential equation for λ whose solutions are subject to the restriction $\lambda > 0$. This is a significant complication relative to scale-independent isotropic hardening in which λ is determined algebraically.

The situation is even more complicated for isotropic materials owing to the fact that in the operative consistency condition, namely

$$F_{\hat{\mathbf{S}}} \cdot (\hat{\mathbf{S}})^{\cdot} + F_{\boldsymbol{\rho}} \cdot \dot{\boldsymbol{\rho}} = 0, \tag{9.45}$$

the derivative $\dot{\boldsymbol{\rho}}$ involves $\dot{\mathbf{M}}$ and its spatial derivatives through the second order, culminating in a second-order partial differential equation for λ.

The formidable obstacles to analysis posed by the differential equations for the plastic multiplier do not arise in the case of rate-sensitive viscoplastic response. For example, (9.17) may be used with F equal to $F(\hat{\mathbf{S}}, \boldsymbol{\alpha})$ or $F(\hat{\mathbf{S}}, \boldsymbol{\rho})$, as appropriate, with vanishing plastic spin in the case of the second alternative. An example of this model, specialized to cubic crystal symmetry with plastic spin incorporated, was proposed and studied by Edmiston et al.

The proposed modifications to the yield function thus model scale effects while retaining intact the structure of conventional viscoplasticity theory. They do not, however, furnish models of the Bauschinger effect. We turn our attention to this problem in the next, and final, section of this book.

9.5 Gradient plasticity

Naturally, there is no requirement that scale effects be confined to the yield function. It is equally permissible to generalize the strain-energy function to take them into account. The resulting models constitute a substantial branch of modern plasticity theory, known collectively as *gradient plasticity*, that has dominated the recent research literature. The book by Gurtin et al. is essential reading in this area.

Gradient plasticity is substantially more complicated than standard plasticity theory. For example, the theory for isotropic materials is complicated by the fact that the Ricci tensor involves spatial derivatives of the plastic strain through the second order. In contrast, the relevant descriptor of scale effects in crystalline materials, the dislocation density, involves the derivatives of the plastic deformation through the first order. Accordingly, we confine ourselves to an outline of a basic framework for crystalline materials and refer to the paper by Steigmann for a discussion of the isotropic theory. To capture the main ideas as simply as possible, we restrict attention to smooth fields.

Thus, we consider strain energies of the form $W(\mathbf{H},\boldsymbol{\alpha})$ in place of functions of the form $W(\mathbf{H})$ discussed previously. The definition (5.20) then yields the strain energy per unit volume of κ as the function

$$\Psi(\mathbf{F},\mathbf{G}, Curl\mathbf{G}) = \mathcal{J}_G W(\mathbf{H},\mathcal{J}_G^{-1}\mathbf{G}Curl\mathbf{G}), \tag{9.46}$$

and the strain energy of an arbitrary part $\pi \subset \kappa$ of the body is

$$\mathcal{U}(\pi, t) = \int_\pi \Psi dV. \tag{9.47}$$

Imposing invariance of W under rotations, as in Chapter 6, we have

$$W(\mathbf{H},\boldsymbol{\alpha}) = U(\hat{\mathbf{E}},\boldsymbol{\alpha}), \tag{9.48}$$

where $\hat{\mathbf{E}}$ is the elastic strain. This is subject to the material symmetry restriction

$$U(\hat{\mathbf{E}},\boldsymbol{\alpha}) = U(\mathbf{R}^t\hat{\mathbf{E}}\mathbf{R},\mathbf{R}^t\boldsymbol{\alpha}\mathbf{R}); \quad \mathbf{R} \in g_{\kappa_i(p)}. \tag{9.49}$$

As in the standard theory, the Piola and Piola–Kirchhoff stresses relative to $\kappa_i(p)$ are $W_{\mathbf{H}}$ and $\hat{\mathbf{S}} = U_{\hat{\mathbf{E}}}$, respectively, and are related by $W_{\mathbf{H}} = \mathbf{H}\hat{\mathbf{S}}$.

9.5.1 Energetic response functions

Consider a one-parameter family of states $\{\mathbf{H}(u), \mathbf{G}(u), \nabla\mathbf{G}(u)\}$ associated with a fixed material point, and let superposed dots denote derivatives with respect to the parameter. Then,

$$\dot{W} = W_{\mathbf{H}} \cdot \dot{\mathbf{H}} + W_{\boldsymbol{\alpha}} \cdot \dot{\boldsymbol{\alpha}}. \tag{9.50}$$

Using (5.3) with the identity $\mathbf{A}\cdot\mathbf{BC} = \mathbf{AC}^t\cdot\mathbf{B} = \mathbf{B}^t\mathbf{A}\cdot\mathbf{C}$, the first term on the right-hand side may be written as

$$W_{\mathbf{H}} \cdot \dot{\mathbf{H}} = W_{\mathbf{H}}\mathbf{K}^t \cdot \dot{\mathbf{F}} + \mathbf{F}^t W_{\mathbf{H}} \cdot \dot{\mathbf{K}}, \tag{9.51}$$

where

$$\dot{\mathbf{K}} = -\mathbf{K}\dot{\mathbf{G}}\mathbf{K}. \tag{9.52}$$

Also, from (9.44) it follows that

$$W_{\boldsymbol{\alpha}} \cdot \dot{\boldsymbol{\alpha}} = (W_{\boldsymbol{\alpha}})\boldsymbol{\alpha}^t\mathbf{K}^t \cdot \dot{\mathbf{G}} - (\dot{J}_G/J_G)W_{\boldsymbol{\alpha}} \cdot \boldsymbol{\alpha} + J_G^{-1}\mathbf{G}^t W_{\boldsymbol{\alpha}} \cdot Curl\dot{\mathbf{G}}. \tag{9.53}$$

Combining these results with

$$\dot{J}_G/J_G = \mathbf{K}^t \cdot \dot{\mathbf{G}}, \tag{9.54}$$

we find the derivative of the referential energy density Ψ to be

$$\dot{\Psi} = J_G[\dot{W} + (\dot{J}_G/J_G)W], \tag{9.55}$$

where

$$\begin{aligned} \dot{W} + (\dot{J}_G/J_G)W \;=\;& W_{\mathbf{H}}\mathbf{K}^t \cdot \dot{\mathbf{F}} + J_G^{-1}\mathbf{G}^t W_{\boldsymbol{\alpha}} \cdot Curl\dot{\mathbf{G}} \\ &+[(W - W_{\boldsymbol{\alpha}} \cdot \boldsymbol{\alpha})\mathbf{I} + (W_{\boldsymbol{\alpha}}\boldsymbol{\alpha}^t - \mathbf{H}^t W_{\mathbf{H}})]\mathbf{K}^t \cdot \dot{\mathbf{G}}. \end{aligned} \tag{9.56}$$

Altogether,

$$\dot{\Psi} = \mathbf{P} \cdot \dot{\mathbf{F}} + J_G\mathbb{E}'\mathbf{K}^t \cdot \dot{\mathbf{G}} + \mathbf{N} \cdot Curl\dot{\mathbf{G}}, \tag{9.57}$$

where

$$\mathbf{P} = \Psi_{\mathbf{F}} = J_G W_{\mathbf{H}}\mathbf{K}^t, \quad \mathbf{N} = \Psi_{Curl\mathbf{G}} = \mathbf{G}^t W_{\boldsymbol{\alpha}} \tag{9.58}$$

and

$$J_G\mathbb{E}'\mathbf{K}^t = \Psi_{\mathbf{G}}, \quad \text{where} \quad \mathbb{E}' = (W - W_{\boldsymbol{\alpha}} \cdot \boldsymbol{\alpha})\mathbf{I} + (W_{\boldsymbol{\alpha}}\boldsymbol{\alpha}^t - \mathbf{H}^t W_{\mathbf{H}}) \tag{9.59}$$

is Eshelby's tensor, referred to $\kappa_i(p)$ and extended to account for the dependence of W on $\boldsymbol{\alpha}$. The tensors $\mathbf{P}$,$\mathbf{N}$, and $\mathbb{E}'$, in which $\mathbf{P}$ is the Piola stress relative to κ, are the relevant response functions in the present theory, and the rate of change of the energy (9.47) is given by

$$\dot{\mathcal{U}}(\pi, t) = \int_{\pi} (\mathbf{P} \cdot \dot{\mathbf{F}} + \mathcal{J}_G \mathbb{E}' \mathbf{K}^t \cdot \dot{\mathbf{G}} + \mathbf{N} \cdot Curl\dot{\mathbf{G}}) dV. \tag{9.60}$$

Problem 9.9. Show that $\mathbb{E}' \rightarrow \mathbf{R}^t \mathbb{E}' \mathbf{R}$ under material symmetry transformations.

9.5.2 Stress power, balance laws, and dissipation

The lesson of Problem 1.2 is that the balance of mechanical energy, Eq. (1.44), follows from the equation of motion, Eq. (1.40). We subsequently showed, conversely, that the equation of motion follows from the *virtual* balance of energy, Eq. (1.50), in which the actual velocity $\mathbf{v} = \dot{\chi}$ is replaced by a kinematically admissible virtual velocity field $\mathbf{u}(\mathbf{x})$. In the case of *equilibrium* ($\dot{\mathbf{v}} = \mathbf{0}$) the latter is equivalent to the statement

$$\mathcal{P} = \mathcal{S} \tag{9.61}$$

for any material region $\pi \subset \kappa$, where

$$\mathcal{P} = \int_{\partial \pi} \mathbf{p} \cdot \mathbf{u} dA + \int_{\pi} \rho_{\kappa} \mathbf{b} \cdot \mathbf{u} dV \tag{9.62}$$

is the virtual power imparted to π and

$$\mathcal{S} = \int_{\pi} \mathbf{P} \cdot \nabla \mathbf{u} dV \tag{9.63}$$

is the virtual stress power. As explained in the book by Gurtin et al. and as is clear from (1.50), Eq. (9.61) covers the non-equilibrium situation, too, if the inertia $\rho_{\kappa}\dot{\mathbf{v}}$ is absorbed into the body force. However, for the sake of discussion we confine attention here to equilibria.

We extend these ideas to gradient plasticity by replacing the "stress power" (1.46) with

$$\mathcal{S}(\pi, t) = \int_{\pi} \{\mathbf{P} \cdot \dot{\mathbf{F}} + \mathcal{J}_G \mathbb{Q} \mathbf{K}^t \cdot \dot{\mathbf{G}} + \mathbf{M} \cdot \nabla \dot{\mathbf{G}}\} dV, \tag{9.64}$$

where $\mathbb{Q}$ and $\mathbf{M}$, respectively, are second- and third-order tensors to be specified. The virtual stress power $\mathcal{S}$ is given by the same expression, but with $\dot{\mathbf{F}} = \nabla \mathbf{u}$ and with $\dot{\mathbf{G}}$ regarded as a virtual plastic velocity field. We intend to integrate the terms involving $\nabla \mathbf{u}$ and $\nabla \dot{\mathbf{G}}$ by parts, using the divergence theorem. The first of these terms is easily handled, as in the passage from (1.50) to (1.51). To treat the second, we write (9.21) in the form

$$\nabla \mathbf{G} = \mathbf{m}_{j|i} \otimes \mathbf{e}^j \otimes \mathbf{e}^i, \quad \text{where} \quad \mathbf{m}_{j|i} = \mathbf{m}_{j,i} - \bar{\Gamma}^k_{ji}\mathbf{m}_k. \tag{9.65}$$

Thus,

$$\nabla \dot{\mathbf{G}} = \dot{\mathbf{m}}_{j|i} \otimes \mathbf{e}^j \otimes \mathbf{e}^i, \tag{9.66}$$

and, writing $\mathbf{M}$ in the form

$$\mathbf{M} = \boldsymbol{\mu}^{kl} \otimes \mathbf{e}_k \otimes \mathbf{e}_l, \tag{9.67}$$

we then have

$$\mathbf{M} \cdot \nabla \dot{\mathbf{G}} = \boldsymbol{\mu}^{ij} \cdot \dot{\mathbf{m}}_{i|j}. \tag{9.68}$$

Problem 9.10. Show that

$$\boldsymbol{\mu}^{ij} \cdot \dot{\mathbf{m}}_{i|j} = (\boldsymbol{\mu}^{ij} \cdot \dot{\mathbf{m}}_i)_{|j} - \boldsymbol{\mu}^{ij}_{|j} \cdot \dot{\mathbf{m}}_i, \quad \text{where} \quad \boldsymbol{\mu}^{ij}_{|j} = \boldsymbol{\mu}^{ij}_{,j} + \boldsymbol{\mu}^{ik}\bar{\Gamma}^j_{kj} + \boldsymbol{\mu}^{kj}\bar{\Gamma}^i_{kj}.$$

The divergence theorem for smooth vector fields $\mathbf{w} = w^i\mathbf{e}_i$ is

$$\int_\pi Div\mathbf{w}dV = \int_{\partial\pi} \mathbf{w} \cdot \boldsymbol{\nu} dA, \tag{9.69}$$

or, in terms of components,

$$\int_\pi w^j_{|j}dV = \int_{\partial\pi} w^j \nu_j dA, \tag{9.70}$$

where $\nu_j = \boldsymbol{\nu} \cdot \mathbf{e}_j$. Applying this with $w^j = \boldsymbol{\mu}^{ij} \cdot \dot{\mathbf{m}}_i$ we then obtain

$$\int_\pi \mathbf{M} \cdot \nabla \dot{\mathbf{G}} dV = \int_{\partial\pi} \boldsymbol{\mu}^{ij}\nu_j \cdot \dot{\mathbf{m}}_i dA - \int_\pi \boldsymbol{\mu}^{ij}_{|j} \cdot \dot{\mathbf{m}}_i dV. \tag{9.71}$$

Using $\dot{\mathbf{m}}_i = \dot{\mathbf{G}}\mathbf{e}_i$, together with the identity $\mathbf{a} \cdot \mathbf{A}\mathbf{b} = \mathbf{a} \otimes \mathbf{b} \cdot \mathbf{A}$, we finally derive

$$\mathcal{S}(\pi, t) = \int_{\partial\pi} (\mathbf{P}\boldsymbol{\nu} \cdot \dot{\boldsymbol{\chi}} + \mathbf{M}\boldsymbol{\nu} \cdot \dot{\mathbf{G}})dA + \int_\pi \{(\mathcal{J}_G \mathbb{Q}\mathbf{K}^t - Div\mathbf{M}) \cdot \dot{\mathbf{G}} - Div\mathbf{P} \cdot \dot{\boldsymbol{\chi}}\}dV, \tag{9.72}$$

where

$$\mathbf{M}\boldsymbol{\nu} = (\boldsymbol{\mu}^{ij} \otimes \mathbf{e}_i)\nu_j \quad \text{and} \quad Div\mathbf{M} = \boldsymbol{\mu}^{ij}_{|j} \otimes \mathbf{e}_i. \tag{9.73}$$

Invoking the virtual power statement (9.61) with $\mathcal{S}$ given by (9.72), in which $\dot{\boldsymbol{\chi}}$ is replaced by $\mathbf{u}$ and $\dot{\mathbf{G}}$ is a virtual field, we conclude that the virtual power has the form

$$\mathcal{P} = \int_{\partial\pi} (\mathbf{p} \cdot \mathbf{u} + \boldsymbol{\Phi} \cdot \dot{\mathbf{G}})dA + \int_{\pi} (\rho_{\kappa}\mathbf{b} \cdot \mathbf{u} + \mathbf{B} \cdot \dot{\mathbf{G}})dV. \tag{9.74}$$

The fields $\boldsymbol{\Phi}$ and $\mathbf{B}$ are force-like quantities that generate power against $\dot{\mathbf{G}}$. Following Di Carlo and Quiligotti, we refer to these as *remodeling forces*.

Because $\mathbf{u}$ and $\dot{\mathbf{G}}$ are independent and arbitrary, we may invoke (9.61) with $\dot{\mathbf{G}} = \mathbf{0}$. This yields (1.50) (with $\dot{\mathbf{v}} = \mathbf{0}$), which, as we have seen, implies that

$$\mathbf{p} = \mathbf{P}\boldsymbol{\nu} \quad \text{on} \quad \partial\pi \quad \text{and} \quad \rho_{\kappa}\mathbf{b} = -Div\mathbf{P} \quad \text{in} \quad \pi. \tag{9.75}$$

The remaining content of (9.61) is

$$\int_{\partial\pi} \boldsymbol{\Phi} \cdot \dot{\mathbf{G}}dA + \int_{\pi} \mathbf{B} \cdot \dot{\mathbf{G}}dV = \int_{\partial\pi} \mathbf{M}\boldsymbol{\nu} \cdot \dot{\mathbf{G}}dA + \int_{\pi} (\mathcal{J}_G\mathbb{Q}\mathbf{K}^t - Div\mathbf{M}) \cdot \dot{\mathbf{G}}dV, \tag{9.76}$$

which, because $\dot{\mathbf{G}}(\mathbf{x})$ is arbitrary, implies that

$$\boldsymbol{\Phi} = \mathbf{M}\boldsymbol{\nu} \quad \text{on} \quad \partial\pi \quad \text{and} \quad \mathbf{B} = \mathcal{J}_G\mathbb{Q}\mathbf{K}^t - Div\mathbf{M} \quad \text{in} \quad \pi. \tag{9.77}$$

These results must be modified if there is any *a priori* restriction on admissible virtual fields $\dot{\mathbf{G}}$. For example, in the case of plastic incompressibility $\dot{\mathbf{G}}$ is restricted by the constraint $tr(\dot{\mathbf{G}}\mathbf{G}^{-1}) = 0$. Here, however, for simplicity's sake we do not impose any constraints of this kind.

Conversely, if (9.75) and (9.77) are satisfied, then

$$\mathcal{P}(\pi, t) = \mathcal{S}(\pi, t), \tag{9.78}$$

where $\mathcal{P}(\pi, t)$ is the *actual* power, given by the right-hand side of (9.74) in which $\mathbf{u}$ is replaced by the actual velocity $\dot{\boldsymbol{\chi}}$ and $\dot{\mathbf{G}}$ is the material time derivative of $\mathbf{G}$.

The dissipation is defined, as in the specialization of (6.56) to equilibrium, by

$$\mathcal{D}(\pi, t) = \mathcal{P}(\pi, t) - \dot{\mathcal{U}}(\pi, t), \tag{9.79}$$

which combines with (9.60), (9.64), and (9.78) to give

$$\mathcal{D}(\pi, t) = \int_{\pi} DdV, \tag{9.80}$$

where

$$D = \mathcal{J}_G(\mathbb{Q} - \mathbb{E}')\mathbf{K}^t \cdot \dot{\mathbf{G}} + \mathbf{M} \cdot \nabla\dot{\mathbf{G}} - \mathbf{N} \cdot Curl\dot{\mathbf{G}}. \tag{9.81}$$

As usual we assume that $\mathcal{D}(\pi, t) \geq 0$ for all $\pi \subset \kappa$ and hence that

$$D \geq 0 \quad \text{at all} \quad \mathbf{x} \in \kappa. \tag{9.82}$$

The expression for D may be reduced by using (5.43) to write $\mathbf{N} \cdot Curl\dot{\mathbf{G}}$ as a linear form in $\nabla\dot{\mathbf{G}}$. Thus,

$$Curl\dot{\mathbf{G}} = \bar{\varepsilon}^{ijk}\mathbf{e}_k \otimes \dot{\mathbf{m}}_{[j,i]} = \bar{\varepsilon}^{ijk}\mathbf{e}_k \otimes \dot{\mathbf{m}}_{[j|i]} = \bar{\varepsilon}^{ijk}\mathbf{e}_k \otimes \dot{\mathbf{m}}_{j|i}, \tag{9.83}$$

where the second equality follows from the symmetry $\bar{\Gamma}^k_{ji} = \bar{\Gamma}^k_{ij}$ and the third from $\bar{\varepsilon}^{ijk}\dot{\mathbf{m}}_{(j|i)} = \mathbf{0}$. We then have

$$\mathbf{N} \cdot Curl\dot{\mathbf{G}} = \bar{\varepsilon}^{ijk}\mathbf{e}_k \cdot \mathbf{N}\dot{\mathbf{m}}_{j|i} = \bar{\varepsilon}^{ijk}\mathbf{N}^t\mathbf{e}_k \cdot \dot{\mathbf{m}}_{j|i} = \boldsymbol{\eta}^{ij} \cdot \dot{\mathbf{m}}_{i|j}, \quad \text{where} \quad \boldsymbol{\eta}^{ij} = \bar{\varepsilon}^{jik}\mathbf{N}^t\mathbf{e}_k, \tag{9.84}$$

and (9.81) may thus be written in the form

$$D = \mathbb{D} \cdot \dot{\mathbf{G}}\mathbf{G}^{-1} + \mathbf{X} \cdot \nabla\dot{\mathbf{G}}, \tag{9.85}$$

with

$$\mathbb{D} = \mathcal{J}_G(\mathbb{Q} - \mathbb{E}') \quad \text{and} \quad \mathbf{X} = \mathbf{M} - \boldsymbol{\eta}^{ij} \otimes \mathbf{e}_i \otimes \mathbf{e}_j. \tag{9.86}$$

Naturally, we require, as in Problem 6.7, that the dissipation be invariant under material symmetry transformations $\mathbf{G} \to \mathbf{R}^t\mathbf{G}$ in which $\mathbf{R} \in g_{\kappa_i(p)}$ is a fixed uniform rotation. To explore the implications, let $\tilde{\mathbb{D}}$ and $\tilde{\mathbf{X}}$ be the transformed values of $\mathbb{D}$ and $\mathbf{X}$. Then, with $\dot{\mathbf{G}}\mathbf{G}^{-1} \to \mathbf{R}^t\dot{\mathbf{G}}\mathbf{G}^{-1}\mathbf{R}$ and $\nabla\dot{\mathbf{G}} \to \mathbf{R}^t\nabla\dot{\mathbf{G}}$, invariance of the dissipation is equivalent to

$$(\mathbf{R}\tilde{\mathbb{D}}\mathbf{R}^t - \mathbb{D}) \cdot \dot{\mathbf{G}}\mathbf{G}^{-1} + (\mathbf{R}\tilde{\mathbf{X}} - \mathbf{X}) \cdot \nabla\dot{\mathbf{G}} = 0. \tag{9.87}$$

As this purports to hold at all material points and for all $\dot{\mathbf{G}}$ and $\nabla\dot{\mathbf{G}}$, and as the latter may be specified independently at any such point, we conclude that

$$\mathbb{D} \to \mathbf{R}^t\mathbb{D}\mathbf{R} \quad \text{and} \quad \mathbf{X} \to \mathbf{R}^t\mathbf{X} \tag{9.88}$$

under symmetry transformations.

Problem 9.11. Show that $\mathbb{Q} \to \mathbf{R}^t\mathbb{Q}\mathbf{R}$ under symmetry transformations.

Problem 9.12. Derive (9.87). Hint: Write $\tilde{\mathbf{X}}$ in the form $\boldsymbol{\xi}^j \otimes \mathbf{e}_j$ and show that $\tilde{\mathbf{X}} \cdot \mathbf{R}^t\nabla\dot{\mathbf{G}} = \boldsymbol{\xi}^j \otimes \mathbf{e}_j \cdot \mathbf{R}^t\dot{\mathbf{G}}_{,i} \otimes \mathbf{e}^i = \boldsymbol{\xi}^i \cdot \mathbf{R}^t\dot{\mathbf{G}}_{,i} = \mathbf{R}\boldsymbol{\xi}^i \cdot \dot{\mathbf{G}}_{,i} = \mathbf{R}\boldsymbol{\xi}^j \otimes \mathbf{e}_j \cdot \dot{\mathbf{G}}_{,i} \otimes \mathbf{e}^i = \mathbf{R}\tilde{\mathbf{X}} \cdot \nabla\dot{\mathbf{G}}$.

Problem 9.13. Show that $\mathbf{N} \to \mathbf{N}\mathbf{R}$ under symmetry transformations and hence that $\mathbf{N} \cdot Curl\dot{\mathbf{G}}$ is invariant. Show that $\boldsymbol{\eta}^{ij} \to \mathbf{R}^t\boldsymbol{\eta}^{ij}$ and hence that $\mathbf{M} \to \mathbf{R}^t\mathbf{M}$.

9.5.3 Example

The foregoing framework is quite general, the only *a priori* restrictions being that the fields $\mathbb{D}$ and $\mathbf{X}$ must be such as to satisfy (9.82) and (9.88). To construct an explicit model within the general framework we must therefore choose particular forms for $\mathbb{D}$ and $\mathbf{X}$, and rely, for the assessment of the resulting model, on *a posteriori* comparisons of its predictions against empirical data. Given the paucity of such data we are content, at present, to illustrate the main ideas as simply as possible. Accordingly, we consider the choices

$$\mathbb{D} = \mu\dot{\mathbf{G}}\mathbf{G}^{-1} \quad \text{and} \quad \mathbf{X} = \mu l^2 \nabla\dot{\mathbf{G}}, \tag{9.89}$$

where μ is a scalar field and l is a material length scale, that is, a material constant having the dimension of length. Assuming μ to be invariant under material symmetry transformations, these trivially satisfy (9.88), reduce (9.85) to

$$D = \mu(\left|\dot{\mathbf{G}}\mathbf{G}^{-1}\right|^2 + l^2\left|\nabla\dot{\mathbf{G}}\right|^2), \tag{9.90}$$

and conform to (9.82) provided that $\mu \geq 0$.

Parallel developments of gradient plasticity discussed in the book by Gurtin et al. are based on models that are "always on," in the sense that no threshold need be met for the onset of plastic flow. Indeed plasticity models of this kind, *sans* yield function, are commonly used to model viscoplastic response. To simplify matters we adopt the same assumption here, and, for the sake of definiteness, choose

$$\mu = 2\eta / P(\hat{\mathbf{E}}, \boldsymbol{\alpha}), \tag{9.91}$$

where η, a positive constant, is a material viscosity and P is a positively valued dimensionless function satisfying the usual material symmetry restriction $P(\hat{\mathbf{E}}, \boldsymbol{\alpha}) = P(\mathbf{R}^t\hat{\mathbf{E}}\mathbf{R}, \mathbf{R}^t\boldsymbol{\alpha}\mathbf{R})$. An example of such a function is

$$P = (1 + pl^2\,|\boldsymbol{\alpha}|^2)^q, \tag{9.92}$$

where p and q are positive dimensionless constants.

According to (9.86) and (9.89), we have

$$\mathcal{J}_G\mathbb{Q} = \mu\dot{\mathbf{G}}\mathbf{G}^{-1} + \mathcal{J}_G\mathbb{E}' \quad \text{and} \quad \mathbf{M} = \mu l^2\nabla\dot{\mathbf{G}} + \boldsymbol{\eta}^{ij} \otimes \mathbf{e}_i \otimes \mathbf{e}_j, \tag{9.93}$$

the latter yielding

$$\boldsymbol{\mu}^{ij} = \mu l^2 e^{ik} e^{jl} \dot{\mathbf{m}}_{k|l} + \boldsymbol{\eta}^{ij}, \tag{9.94}$$

and the balance law $(9.77)_2$ reduces to

$$\dot{\mathbf{G}}\mathbf{G}^{-1}\mathbf{K}^t = \mu^{-1}(\mathbf{B} - \mathcal{J}_G\mathbb{E}'\mathbf{K}^t + Div\mathbf{M}). \tag{9.95}$$

We observe, from (9.73), (9.91), and (9.93), that the highest-order spatial derivative of $\dot{\mathbf{G}}$ occurring in this equation is $l^2 e^{ik} e^{jl} \dot{\mathbf{m}}_{k|lj} \otimes \mathbf{e}_i$. Using $\dot{\mathbf{m}}_k = \dot{\mathbf{G}}\mathbf{e}_k$, and the consequent fact that $\dot{\mathbf{m}}_{k|lj} = \dot{\mathbf{G}}_{|lj}\mathbf{e}_k$, we reduce this to

$$l^2 e^{ik} e^{jl} \dot{\mathbf{m}}_{k|lj} \otimes \mathbf{e}_i = l^2 (e^{jl} \dot{\mathbf{G}}_{|lj}) \mathbf{e}^i \otimes \mathbf{e}_i = l^2 e^{jl} \dot{\mathbf{G}}_{|lj} = l^2 \Delta \dot{\mathbf{G}}, \tag{9.96}$$

where $\Delta(\cdot) = e^{ij}(\cdot)_{|ij}$ is the referential Laplacian. Accordingly, (9.95) is a quasilinear second-order elliptic partial differential equation for $\dot{\mathbf{G}}$. It becomes a linear equation if we take $P = const.$ in (9.91).

If no external agency is available to supply a volumetric remodeling force, i.e., if $\mathbf{B}$ vanishes, then (9.95) furnishes the flow rule

$$\dot{\mathbf{G}}\mathbf{G}^{-1} = \mu^{-1}\{(Div\mathbf{M})\mathbf{G}^t - J_G \mathbb{E}'\}, \tag{9.97}$$

which, together with (9.91), may be compared to the first branch of (9.17). It may also be compared to conventional scale-independent models of *kinematic hardening*, in which the bracket on the right-hand side of (9.97) is replaced by a difference between the stress $\hat{\mathbf{S}}$ and a tensor-valued variable that purports to describe the Bauschinger effect. See Eq. (3.1.9) in Lubliner's book, for example. With reference to (9.59), written in the form

$$\mathbb{E}' = (U - U_\alpha \cdot \alpha)\mathbf{I} + (U_\alpha \alpha^t - \mathbf{H}^t\mathbf{H}\hat{\mathbf{S}}), \tag{9.98}$$

the tensor variable in question is seen to be represented, in the case of small elastic strain ($\mathbf{H}^t\mathbf{H} \simeq \mathbf{I}$), by the energetic effect of the dislocation density together with the divergence term in (9.97). Solutions to (9.97) also exhibit plastic spin. The Bauschinger effect, associated with the emergence of dislocations and hence an inherently scale-dependent phenomenon, is thus a natural outcome of gradient plasticity theory for crystalline materials.

Equation (9.97) requires the specification of boundary conditions on $\partial\kappa$. With reference to (9.76) these are seen to be consistent with the specification of $\mathbf{\Phi}$ or $\mathbf{G}$. For example, taking $\mathbf{G}$ to be fixed ($\dot{\mathbf{G}} = \mathbf{0}$) on a part of $\partial\kappa$ and $\mathbf{\Phi} = \mathbf{0}$ on the remainder corresponds to the "hard" and "free" conditions specified by Gurtin et al., the former simulating a boundary, abutting a hard material, that blocks the passage of dislocations via plastic slip, and the latter a boundary across which they can flow freely. Given an initial field $\mathbf{G}(\mathbf{x}, t_0)$ compatible with the condition imposed on a hard part of the boundary, the solution to the boundary-value problem for $\dot{\mathbf{G}}(\mathbf{x}, t)$ can then be used to update $\mathbf{G}(\mathbf{x}, t)$.

Of course the tentative model offered in this subsection is merely one among an infinite number of possibilities encompassed by the general framework. Ultimately, the usefulness of this or any similar model must be judged by the community of plasticians.

References

Atai, A. A., and Steigmann, D. J. (2014). Transient elastic-viscoplastic dynamics of thin sheets. *J. Mech. Mat. Struct.* 9, 557–74.

Batchelor, G. K. (Ed.) (1958). *The Scientific Papers of Sir Geoffrey Ingram Taylor*, Vol. 1: *Mechanics of Solids*. Cambridge University Press, Cambridge, UK.

Cermelli, P., and Gurtin, M. E. (2001). On the characterization of geometrically necessary dislocations in finite plasticity. *J. Mech. Phys. Solids* 49, 1539–68.

Di Carlo, A., and Quiligotti, S. (2002). Growth and balance. *Mech. Res. Comm.* 29, 449–456.

Edmiston, J., Steigmann, D. J., Johnson, G., and Barton, N. (2013). A model for elastic-viscoplastic deformations of crystalline solids based on material symmetry: Theory and plane-strain simulations. *Int. J. Eng. Sci.* 63, 10–22.

Fleck, N. A., Muller, G. N., Ashby, M. F., et al. (1994). Strain gradient plasticity: Theory and experiment. *Acta Metall. et Mater.* 42, 475–87.

Gurtin, M. E., Fried, E., and Anand, L. (2010). *Mechanics and Thermodynamics of Continua*. Cambridge University Press, Cambridge, UK.

Hill, R. (1950). *The Mathematical Theory of Plasticity*. Clarendon Press, Oxford.

Hutchinson, J.W. (2000). Plasticity at the micron scale. *Int. J. Solids Structures* 37, 225–38.

Krishnan, J., and Steigmann, D. J. (2014). A polyconvex formulation of isotropic elastoplasticity theory. *IMA J. Appl. Maths.* 79, 722–38.

Lubliner, J. (2008). *Plasticity Theory*. Dover, New York.

Perzyna, P. (1962/3). The constitutive equations for rate-sensitive plastic materials. *Quart. Appl. Math.* 20, 321–32.

Steigmann, D. J. (2022). Gradient plasticity in isotropic solids. *Math. Mech. Solids.* 27, 1896–1912.

Stölken, J. S., and Evans, A. G. (1998). A microbend test method for measuring the plasticity length scale. *Acta Mater.* 46, 5109–15.

Taylor, G.I. (1934). The mechanism of plastic deformation of crystals. Part 1: Theoretical. *Proc. R. Soc. Lond.* **A 145**, 326–87.

Teodosiu, C. (Ed.) (1997). *Large Plastic Deformation of Crystalline Aggregates*. CISM Courses and Lectures, Vol. 376. Springer, Vienna.

Solutions to selected problems

2.1. If $\bar{\mathbf{C}} = \mathbf{C}$ then $\mathbf{F}^t\bar{\mathbf{Q}}^t\bar{\mathbf{Q}}\mathbf{F} = \mathbf{F}^t\mathbf{F}$. Pre-multiply by $\mathbf{F}^{-t}$ and post-multiply by $\mathbf{F}^{-1}$ to conclude that $\bar{\mathbf{Q}}^t\bar{\mathbf{Q}} = \mathbf{i}$. Conversely, if $\bar{\mathbf{Q}}^t\bar{\mathbf{Q}} = \mathbf{i}$ then it follows immediately that $\bar{\mathbf{C}} = \mathbf{C}$.

2.3. $\mathbf{AB} \cdot \mathbf{D} = tr(\mathbf{ABD}^t) = tr((\mathbf{ABD}^t)^t) = tr(\mathbf{DB}^t\mathbf{A}^t) = \mathbf{DB}^t \cdot \mathbf{A}$. Also, $\mathbf{A} \cdot \mathbf{BD} = tr(\mathbf{A}(\mathbf{BD})^t) = tr(\mathbf{AD}^t\mathbf{B}^t) = tr(\mathbf{B}^t\mathbf{AD}^t) = \mathbf{B}^t\mathbf{A} \cdot \mathbf{D}$.

2.5(b). Suppose a uniform pressure p is assigned on the entire boundary. We have

$$\begin{aligned}\mathcal{P} &= -p\int_{\partial\kappa_t} \mathbf{n}\cdot\dot{\chi}da = -p\int_{\partial\kappa} \mathbf{F}^*\nu\cdot\dot{\chi}dA \\ &= -p\int_{\kappa} Div[(\mathbf{F}^*)^t\dot{\chi}]dV = -p\int_{\kappa}(\dot{\chi}\cdot Div\mathbf{F}^* + \mathbf{F}^*\cdot\dot{\mathbf{F}})dV = -p\int_{\kappa}\dot{\mathcal{J}}_F dV,\end{aligned}$$

where we've used the Piola identity $Div\mathbf{F}^* = \mathbf{0}$ together with $\nabla\dot{\chi} = \dot{\mathbf{F}}$ and $(\mathcal{J}_F)_{\mathbf{F}} = \mathbf{F}^*$. If p is fixed, independent of time, then

$$\mathcal{P} = \tfrac{d}{dt}\mathcal{L},$$

where

$$\mathcal{L} = -p\int_{\kappa}\mathcal{J}_F dV = -p\,vol(\kappa_t).$$

A simpler way to arrive at the same conclusion is to use

$$\int_{\partial\kappa_t}\mathbf{n}\cdot\dot{\chi}da = \int_{\kappa_t} div\dot{\chi}dv = \int_{\kappa_t}(\dot{\mathcal{J}}_F/\mathcal{J}_F)dv = \int_{\kappa}\dot{\mathcal{J}}_F dV.$$

This is not the only kind of pressure loading for which a load potential exists. Further examples include a uniform volume-dependent pressure and a hydrostatic pressure in a uniform gravitational field. Can you derive the forms of the associated load potentials?

2.6. The mechanical power identity for the entire body is $\mathcal{P} = \mathcal{S} + \frac{d}{dt}\mathcal{K}$, where

$$\mathcal{S} = \int_{\kappa}\mathbf{P}\cdot\dot{\mathbf{F}}dV.$$

Using

$$\mathbf{P} = \Psi_{\mathbf{F}} + \mu\mathbf{F}\dot{\mathbf{E}},$$

where μ is the viscosity and $\dot{\mathbf{E}}$ is the material derivative of the Lagrange strain, we have

$$\mathbf{P} \cdot \dot{\mathbf{F}} = \dot{\Psi} + \mu \mathbf{F}\dot{\mathbf{E}} \cdot \dot{\mathbf{F}},$$

where (noting that $\mathbf{A} \cdot \mathbf{BD} = \mathbf{B}^t\mathbf{A} \cdot \mathbf{D}$)

$$\mathbf{F}\dot{\mathbf{E}} \cdot \dot{\mathbf{F}} = \dot{\mathbf{F}}^t\mathbf{F} \cdot \dot{\mathbf{E}} = Sym(\dot{\mathbf{F}}^t\mathbf{F}) \cdot \dot{\mathbf{E}} = \dot{\mathbf{E}} \cdot \dot{\mathbf{E}}.$$

We have used $\mathbf{E} = \frac{1}{2}(\mathbf{F}^t\mathbf{F} - \mathbf{I})$ together with the fact that $\dot{\mathbf{E}}$ is symmetric. Thus,

$$\mathbf{P} \cdot \dot{\mathbf{F}} = \dot{\Psi} + \mu \left|\dot{\mathbf{E}}\right|^2$$

and

$$\mathcal{S} = \tfrac{d}{dt}\mathcal{U} + \int_\kappa \mu \left|\dot{\mathbf{E}}\right|^2 dV, \quad \text{where} \quad \mathcal{U} = \int_\kappa \Psi dV.$$

The *dissipation* $\mathcal{D}$ is

$$\mathcal{D} = \mathcal{P} - \tfrac{d}{dt}(\mathcal{U} + \mathcal{K}) = \mathcal{S} - \tfrac{d}{dt}\mathcal{U} = \int_\kappa \mu \left|\dot{\mathbf{E}}\right|^2 dV,$$

where we have invoked the mechanical energy balance in the second equality. Thus, $\mathcal{D}$ vanishes in the case of purely elastic response ($\mu = 0$).

(a) Assuming $\mu > 0$ we immediately conclude that $\mathcal{D} \geq 0$. If $\dot{\mathbf{E}} = \mathbf{0}$ at all material points, then obviously $\mathcal{D} = 0$. Conversely, if $\mathcal{D} = 0$, then $\dot{\mathbf{E}} = \mathbf{0}$ at all points. For if there is a point where $\dot{\mathbf{E}} \neq \mathbf{0}$, then by continuity there is a neighborhood of that point where $|\dot{\mathbf{E}}| > 0$, yielding a positive value of the integral and hence $\mathcal{D} > 0$, a contradiction.

(b) In the case of conservative loading we have $\mathcal{P} = \frac{d}{dt}\mathcal{L}$ and $E = (\mathcal{U} - \mathcal{L}) + \mathcal{K}$. Then, $\mathcal{D} = -\frac{d}{dt}E$ and the inequality $\mathcal{D} \geq 0$ is equivalent to

$$\tfrac{d}{dt}E \leq 0.$$

2.7. For the first part, write $\mathbf{x} = \mathbf{x}_0 + \epsilon\mathbf{w}$ in (2.60), with $dV(\mathbf{x}) = \epsilon^3 dV(\mathbf{w})$. Divide by $\epsilon^3 (> 0)$ and let $\epsilon \to 0$ to get (2.63). In the second part, the coefficient of i is found to be

$$\nabla\xi_2 \cdot \mathcal{A}[\nabla\xi_1] - \nabla\xi_1 \cdot \mathcal{A}[\nabla\xi_2] = \nabla\xi_1 \cdot \mathcal{A}^t[\nabla\xi_2] - \nabla\xi_1 \cdot \mathcal{A}[\nabla\xi_2],$$

which vanishes by the major symmetry of $\mathcal{A}$ ($\mathcal{A}^t = \mathcal{A}$).

2.9. (a) If $g_{\mathcal{K}_1(p)} = Orth^+$, then $\mathbf{R} \in g_{\mathcal{K}_1(p)}$ is an arbitrary rotation. The corresponding element of $g_{\mathcal{K}_2(p)}$ is $\mathbf{KRK}^{-1}$, where

$$\mathbf{K} = \lambda\mathbf{k} \otimes \mathbf{k} + \lambda^{-1/2}(\mathbf{I} - \mathbf{k} \otimes \mathbf{k}) \quad \text{and} \quad \mathbf{K}^{-1} = \lambda^{-1}\mathbf{k} \otimes \mathbf{k} + \lambda^{1/2}(\mathbf{I} - \mathbf{k} \otimes \mathbf{k}).$$

Assume the material to be isotropic relative to κ_2, so that $g_{\kappa_2(p)} = Orth^+$. Then $(\mathbf{KRK}^{-1})^{-1} = (\mathbf{KRK}^{-1})^t$, which is equivalent to $\mathbf{K}^2\mathbf{R} = \mathbf{RK}^2$ because $\mathbf{K}$ is symmetric. We obtain

$$\mathbf{K}^2\mathbf{R} - \mathbf{RK}^2 = (\lambda^2 - \lambda^{-1})(\mathbf{k} \otimes \mathbf{R}^t\mathbf{k} - \mathbf{Rk} \otimes \mathbf{k}).$$

There are infinitely many rotations $\mathbf{R}$ for which this is not zero, unless $\lambda = 1$. For such $\mathbf{R}$ we have $\mathbf{KRK}^{-1} \notin Orth^+$ and so our assumption is false: $g_{\kappa_2(p)} \neq Orth^+$ and the material is not isotropic relative to κ_2 unless $\lambda = 1$, i.e., unless $\mathbf{K} = \mathbf{I}$.

(b) If $\mathbf{R} \in Orth^+$ and $\mathbf{Rk} = \mathbf{k}$, then $\mathbf{R}^t\mathbf{k} = \mathbf{k}$ and $\mathbf{K}^2\mathbf{R} - \mathbf{RK}^2 = \mathbf{0}$; thus, $\mathbf{KRK}^{-1} \in Orth^+$ from part (a). Using Rodrigues' formula we have

$$\mathbf{R} = \mathbf{k} \otimes \mathbf{k} + \mathbf{S}, \quad \text{where} \quad \mathbf{S} = \cos\theta(\mathbf{i} \otimes \mathbf{i} + \mathbf{j} \otimes \mathbf{j}) + \sin\theta(\mathbf{j} \otimes \mathbf{i} - \mathbf{i} \otimes \mathbf{j}),$$

in which θ is an arbitrary real number and $\{\mathbf{i}, \mathbf{j}, \mathbf{k}\}$ is a right-handed orthonormal basis. We obtain

$$\mathbf{KRK}^{-1} = \lambda^{-1}\mathbf{Kk} \otimes \mathbf{k} + \lambda^{1/2}\mathbf{KS} = \mathbf{k} \otimes \mathbf{k} + \mathbf{S} = \mathbf{R},$$

and the material is also transversely isotropic relative to κ_2.

3.6. $d\mathbf{y} = \mathbf{g}_i d\xi^i = \mathbf{i}_1 d\xi^1 + (\sin\phi\mathbf{i}_1 + \cos\phi\mathbf{i}_2)d\xi^2 + \mathbf{i}_3 d\xi^3 = \mathbf{i}_i dy^i$, yielding $dy^1 = d\xi^1 + (\sin\phi)d\xi^2$, $dy^2 = (\cos\phi)d\xi^2$ and $dy^3 = d\xi^3$. Integrate with ϕ = constant and set $y^i(0,0,0) = 0$ to get $y^1 = \xi^1 + (\sin\phi)\xi^2$, $y^2 = (\cos\phi)\xi^2$ and $y^3 = \xi^3$. The rest is straightforward.

3.7. Use $\mathbf{A} = A^i_{\cdot j}\mathbf{g}_i \otimes \mathbf{g}^j$ with $\mathbf{b} \cdot \mathbf{A}^t\mathbf{a} = \mathbf{a} \cdot \mathbf{Ab}$ for all $\mathbf{a}, \mathbf{b}$ to get $\mathbf{g}_k \cdot \mathbf{A}^t\mathbf{g}^l = \mathbf{g}^l \cdot (A^i_{\cdot j}\mathbf{g}_i \otimes \mathbf{g}^j)\mathbf{g}_k = A^i_{\cdot j}\delta^j_k\mathbf{g}^l \cdot \mathbf{g}_i = A^i_{\cdot j}\delta^j_k\delta^l_i = A^l_{\cdot k}$. Then, $\mathbf{A}^t = (\mathbf{g}_k \cdot \mathbf{A}^t\mathbf{g}^l)\mathbf{g}^k \otimes \mathbf{g}_l = A^l_{\cdot k}\mathbf{g}^k \otimes \mathbf{g}_l = A_j^{\cdot i}\mathbf{g}_i \otimes \mathbf{g}^j$ where, in the last step, we have raised and lowed indices using the metric and reciprocal metric. Comparing with $\mathbf{A}$ we have that $A_j^{\cdot i} = A^i_{\cdot j}$ if $\mathbf{A}^t = \mathbf{A}$ and $A_j^{\cdot i} = -A^i_{\cdot j}$ if $\mathbf{A}^t = -\mathbf{A}$. The rest is straightforward.

3.13. We need $d\mathbf{y} = \mathbf{g}_i d\xi^i = \bar{\mathbf{e}}_r d\bar{r} + \bar{r} d\bar{\mathbf{e}}_r + \mathbf{i}_3 d\bar{z}$, where $d\bar{\mathbf{e}}_r = \cos\phi\mathbf{e}_\theta(\theta)d\theta$ and $\{\xi^i\} = \{\bar{r}, \theta, \bar{z}\}$, yielding $\mathbf{g}_1 = \bar{\mathbf{e}}_r$, $\mathbf{g}_2 = \bar{r}\cos\phi\mathbf{e}_\theta$ and $\mathbf{g}_3 = \mathbf{i}_3$. The rest is straightforward.

3.15. We have $\mathbf{g}_1 = \mathbf{y}_{,r} = \cos\theta\mathbf{n} + \sin\theta\mathbf{b}$, $\mathbf{g}_2 = \mathbf{y}_{,\theta} = -r\sin\theta\mathbf{n} + r\cos\theta\mathbf{b}$, $\mathbf{g}_3 = \mathbf{y}_{,s} = \mathbf{r}' + r\cos\theta\mathbf{n}' + r\sin\theta\mathbf{b}' = (1 - \kappa r\cos\theta)\mathbf{t} + \tau r\cos\theta\mathbf{b} - \tau r\sin\theta\mathbf{n}$. Then, $g_{11} = 1, g_{22} = r^2, g_{23} = g_{32} = \tau r^2$, and $g_{33} = H$, all other g_{ij} being zero, where $H = (1 - \kappa r\cos\theta)^2 + \tau^2 r^2$. This yields $g = \det(g_{ij}) = r^2(1 - \kappa r\cos\theta)$. Also, $|\kappa r| < 1$ and $|\cos\theta| \leq 1$ so $\sqrt{g} = r(1 - \kappa r\cos\theta)$. We also get $g^{11} = 1, g^{22} = H/g, g^{33} = r^2/g$, $g^{23} = g^{32} = -\tau r^2/g$, all others zero. For a function $f(r, \theta)$ we use $\Delta f = g^{-1/2}(g^{1/2}g^{ij}f_{,j})_{,i}$ to get (eventually!)

$$\begin{aligned}\Delta f = {} & \frac{\partial^2 f}{\partial r^2} + \left(1 - \frac{\kappa r\cos\theta}{1 - \kappa r\cos\theta}\right)\frac{1}{r}\frac{\partial f}{\partial r} + \left(\frac{\kappa\sin\theta}{1 - \kappa r\cos\theta}\right)\left[1 - \frac{\tau^2 r^2}{(1 - \kappa r\cos\theta)^2}\right]\frac{1}{r}\frac{\partial f}{\partial\theta} \\ & + \left[1 + \frac{\tau^2 r^2}{(1 - \kappa r\cos\theta)^2}\right]\frac{1}{r^2}\frac{\partial^2 f}{\partial\theta^2}.\end{aligned}$$

3.16. The transformation formula

$$\bar{A}^k_{\cdot ij} = \frac{\partial\bar{\xi}^k}{\partial\xi^l}\frac{\partial\xi^m}{\partial\bar{\xi}^i}\frac{\partial\xi^n}{\partial\bar{\xi}^j}A^l_{\cdot mn},$$

for the components $A^k_{.ij}$ of a third-order tensor, follows directly from $A^k_{.ij}\mathbf{g}_k \otimes \mathbf{g}^i \otimes \mathbf{g}^j = \mathbf{A} = \bar{A}^k_{.ij}\bar{\mathbf{g}}_k \otimes \bar{\mathbf{g}}^i \otimes \bar{\mathbf{g}}^j$.

(a) Using (3.56) we have

$$\begin{aligned}\Gamma^k_{ij} &= \mathbf{g}^k \cdot \mathbf{g}_{i,j} = \frac{\partial \xi^k}{\partial \bar{\xi}^l}\bar{\mathbf{g}}^l \cdot \frac{\partial \mathbf{g}_i}{\partial \bar{\xi}^m}\frac{\partial \bar{\xi}^m}{\partial \xi^j} \\ &= \frac{\partial \xi^k}{\partial \bar{\xi}^l}\frac{\partial \bar{\xi}^m}{\partial \xi^j}\bar{\mathbf{g}}^l \cdot \frac{\partial}{\partial \bar{\xi}^m}\left(\frac{\partial \bar{\xi}^n}{\partial \xi^i}\bar{\mathbf{g}}_n\right) \\ &= \frac{\partial \xi^k}{\partial \bar{\xi}^l}\frac{\partial \bar{\xi}^m}{\partial \xi^j}\frac{\partial \bar{\xi}^n}{\partial \xi^i}\bar{\Gamma}^l_{nm} + \frac{\partial \xi^k}{\partial \bar{\xi}^l}\frac{\partial \bar{\xi}^m}{\partial \xi^j}\frac{\partial}{\partial \bar{\xi}^m}\left(\frac{\partial \bar{\xi}^l}{\partial \xi^i}\right),\end{aligned}$$

where $\frac{\partial}{\partial \bar{\xi}^m}\left(\frac{\partial \bar{\xi}^l}{\partial \xi^i}\right) = \frac{\partial}{\partial \xi^s}\left(\frac{\partial \bar{\xi}^l}{\partial \xi^i}\right)\frac{\partial \xi^s}{\partial \bar{\xi}^m} = \frac{\partial \xi^s}{\partial \bar{\xi}^m}\frac{\partial^2 \bar{\xi}^l}{\partial \xi^s \partial \xi^i}$. Use this with $\frac{\partial \xi^s}{\partial \bar{\xi}^m}\frac{\partial \bar{\xi}^m}{\partial \xi^j} = \delta^s_j$ to get

$$\Gamma^k_{ij} = \frac{\partial \xi^k}{\partial \bar{\xi}^l}\frac{\partial \bar{\xi}^m}{\partial \xi^j}\frac{\partial \bar{\xi}^n}{\partial \xi^i}\bar{\Gamma}^l_{nm} + \frac{\partial \xi^k}{\partial \bar{\xi}^l}\frac{\partial^2 \bar{\xi}^l}{\partial \xi^i \partial \xi^j}.$$

The presence of the second term on the right implies that Γ^k_{ij} are not the components of a tensor.

(b) We have

$$\begin{aligned}\Gamma^k_{ji} &= \frac{\partial \xi^k}{\partial \bar{\xi}^l}\frac{\partial \bar{\xi}^m}{\partial \xi^i}\frac{\partial \bar{\xi}^n}{\partial \xi^j}\bar{\Gamma}^l_{nm} + \frac{\partial \xi^k}{\partial \bar{\xi}^l}\frac{\partial^2 \bar{\xi}^l}{\partial \xi^j \partial \xi^i} \\ &= \frac{\partial \xi^k}{\partial \bar{\xi}^l}\frac{\partial \bar{\xi}^n}{\partial \xi^i}\frac{\partial \bar{\xi}^m}{\partial \xi^j}\bar{\Gamma}^l_{mn} + \frac{\partial \xi^k}{\partial \bar{\xi}^l}\frac{\partial^2 \bar{\xi}^l}{\partial \xi^i \partial \xi^j},\end{aligned}$$

implying that

$$\Gamma^k_{ij} - \Gamma^k_{ji} = \frac{\partial \xi^k}{\partial \bar{\xi}^l}\frac{\partial \bar{\xi}^n}{\partial \xi^i}\frac{\partial \bar{\xi}^m}{\partial \xi^j}(\bar{\Gamma}^l_{nm} - \bar{\Gamma}^l_{mn}).$$

Thus, $\Gamma^k_{[ij]}$ is a third-order tensor.

3.19. (a) Let $A_{ij} = v_{i;j} = v_{i,j} - v_m\Gamma^m_{ij}$. Then,

$$\begin{aligned}A_{ij;k} &= A_{ij,k} - A_{lj}\Gamma^l_{ik} - A_{il}\Gamma^l_{jk} \\ &= v_{i,jk} - v_{m,j}\Gamma^m_{ik} - v_m\Gamma^m_{ik,j} - v_{l,k}\Gamma^l_{ij} + v_m\Gamma^m_{lk}\Gamma^l_{ij} - v_{i,l}\Gamma^l_{kj} + v_m\Gamma^m_{il}\Gamma^l_{kj} \\ &\quad + v_{l,k}\Gamma^l_{ij} - v_{m,k}\Gamma^m_{ij} - v_{l,j}\Gamma^l_{ik} + v_{m,j}\Gamma^m_{ik},\end{aligned}$$

in which the last line is seen to vanish identically. Thus,

$$v_{i;jk} - v_{i;kj} = (\Gamma^m_{ik,j} - \Gamma^m_{ij,k} + \Gamma^m_{lj}\Gamma^l_{ik} - \Gamma^m_{lk}\Gamma^l_{ij})v_m - (v_{i,l} - v_m\Gamma^m_{il})(\Gamma^l_{jk} - \Gamma^l_{kj}),$$

which is the stated formula.

(c) Let ξ^i be general coordinates and let $\bar{\xi}^i = y^i$ be Cartesian coordinates. The latter are admissible coordinates in Euclidean space. Because $v_{i;jk}$ are the covariant components of a third-order tensor, we have

$$v_{i;jk} = \frac{\partial y^l}{\partial \xi^i}\frac{\partial y^m}{\partial \xi^j}\frac{\partial y^n}{\partial \xi^k}\bar{v}_{l,mn},$$

where, on the right-hand side, the partial derivatives of $\bar{v}_l$ are with respect to the y' s and we have used the fact that the $\bar{\Gamma}^l_{ik}$ are identically zero. Then,

$$v_{i;kj} = \frac{\partial y^l}{\partial \xi^i}\frac{\partial y^m}{\partial \xi^k}\frac{\partial y^n}{\partial \xi^j}\bar{v}_{l,mn} = \frac{\partial y^l}{\partial \xi^i}\frac{\partial y^m}{\partial \xi^j}\frac{\partial y^n}{\partial \xi^k}\bar{v}_{l,nm},$$

and

$$v_{i;jk} - v_{i;kj} = \frac{\partial y^l}{\partial \xi^i}\frac{\partial y^m}{\partial \xi^j}\frac{\partial y^n}{\partial \xi^k}(\bar{v}_{l,mn} - \bar{v}_{l,nm}) = 0$$

if $\bar{v}_l$ is twice continuously differentiable.

4.8. Symmetry of the Ricci tensor follows directly by using the symmetry of the metric together with the major symmetry and minor skew symmetries of the Riemann tensor. Thus, $R^{ij} = m_{kl}R^{kjil} = m_{kl}R^{ilkj} = m_{lk}R^{lijk} = R^{ji}$.

4.10. We present the conventional derivation based on the Riemann tensor. We have

$$R_{pmlj} = \Gamma_{pmj,l} - \Gamma_{pml,j} + \Gamma^i_{ml}\Gamma_{ipj} - \Gamma^i_{mj}\Gamma_{ipl},$$

where

$$\Gamma_{kij} = \tfrac{1}{2}(g_{ki,j} + g_{jk,i} - g_{ij,k}) \quad \text{and} \quad \Gamma^l_{ij} = g^{lk}\Gamma_{kij}.$$

Using

$$(e_{ij} + 2E_{ij})_{|k} = g_{ij|k} = g_{ij,k} - g_{lj}\bar{\Gamma}^l_{ik} - g_{li}\bar{\Gamma}^l_{jk},$$

together with $e_{ij|k} = 0$ and $\Gamma^l_{ij} = \Gamma^l_{ji}$, we get

$$\Gamma_{kij} = \bar{\Gamma}_{kij} + 2\gamma_{kij},$$

where

$$\gamma_{kij} = E_{kl}\bar{\Gamma}^l_{ij} + E_{kij},$$

with

$$E_{kij} = E_{ki|j} + E_{jk|i} - E_{ij|k}. \qquad (\star)$$

Choosing the convected coordinates to be Cartesian in the reference configuration we have $\bar{\Gamma}^l_{ij} = 0$, identically, and hence the simplification

$$\Gamma_{kij} = 2E_{kij}, \quad \text{with} \quad E_{kij} = E_{ki,j} + E_{jk,i} - E_{ij,k},$$

and

$$\Gamma^l_{ij} = 2g^{lk}E_{kij}.$$

Substituting into the expression for R, we get

$$R_{pmlj} = 2[(E_{pmj,l} - E_{pml,j}) + 2g^{ik}(E_{kml}E_{ipj} - E_{kmj}E_{ipl})],$$

where

$$(g^{ik}) = (e_{ij} + 2E_{ij})^{-1}.$$

Using $R_{pmlj} = 0$ and transforming to arbitrary coordinates, the compatibility conditions reduce to

$$E_{pmj|l} - E_{pml|j} + 2g^{ik}(E_{kml}E_{ipj} - E_{kmj}E_{ipl}) = 0,$$

where E_{kml} is given by $(\star)$, and where

$$E_{ijk|l} = E_{ijk,l} - E_{mjk}\bar{\Gamma}^m_{il} - E_{imk}\bar{\Gamma}^m_{jl} - E_{ijm}\bar{\Gamma}^m_{kl}.$$

Using (4.36) with the expression for R_{pmlj} yields the six non-trivial compatibility conditions.

4.12. We have $\dot{\mathbf{E}} = \frac{1}{2}\dot{g}_{ij}\mathbf{e}^i \otimes \mathbf{e}^j$ and $\dot{W} = \dot{U} = U_{\mathbf{E}} \cdot \dot{\mathbf{E}} = \mathbf{S} \cdot \dot{\mathbf{E}} = \frac{1}{2}\mathbf{S} \cdot \mathbf{e}^i \otimes \mathbf{e}^j \dot{g}_{ij} = \frac{1}{2}S^{ij}\dot{g}_{ij} = \frac{1}{2}\mathcal{J}_F T^{ij}\dot{g}_{ij}$. The result follows by equating this to $\dot{W} = \frac{\partial W}{\partial g_{ij}}\dot{g}_{ij}$ and demanding that the two expressions agree for all symmetric $\dot{g}_{ij}$. Note that for the partial-derivative notation to be meaningful, we must replace g_{ij} by $\frac{1}{2}(g_{ij} + g_{ji})$ in the expression for W and regard g_{ij} as being non-symmetric, so that $\frac{\partial W}{\partial g_{ij}} = \frac{\partial W}{\partial g_{ji}}$.

5.2. $\mathbf{H}^{-1}d\mathbf{y} = \mathbf{H}^{-1}\mathbf{F}d\mathbf{x} = (\mathbf{FK})^{-1}\mathbf{F}d\mathbf{x} = \mathbf{K}^{-1}\mathbf{F}^{-1}\mathbf{F}d\mathbf{x} = \mathbf{G}d\mathbf{x}$.

5.4. From the referential counterparts of (3.85) and (3.117), we have $(Curl\mathbf{A})^t = B^{jm}\mathbf{e}_j \otimes \mathbf{e}_m$, where $B^{jm} = \bar{\varepsilon}^{ilm}A^j_{\cdot l|i}$; and, $Div[(Curl\mathbf{A})^t] = B^{jm}_{|m}\mathbf{e}_j$, where $B^{jm}_{|m} = \bar{\varepsilon}^{mil}A^j_{\cdot l|im}$ in which use has been made of the fact that the permutation tensor is covariantly constant. $B^{jm}_{|m}$ vanishes because, in Euclidean space, $A^j_{\cdot l|im}$ is symmetric in the last pair of subscripts. This follows from the fact that $A^j_{\cdot l|im}$ are the components of a fourth-order tensor, the covariant derivatives reducing, in a Cartesian system, to symmetric partial derivatives. It is a simple matter to show that this symmetry is preserved by the relevant tensor transformation formula.

(b) From (5.43) it follows that

$$(Curl\mathbf{G})^t = \mathbf{L}^k \otimes \mathbf{e}_k,$$

where

$$\mathbf{L}^k = \bar{\epsilon}^{ijk} \hat{T}^l_{ji} \mathbf{m}_l.$$

Then, from (4.51), (4.54), (5.29), (5.31), and $\bar{\epsilon}^{ijk} = e^{ijk}/\sqrt{e}$ (see (3.32)),

$$0 = Div[(Curl\mathbf{G})^t] = \tfrac{1}{\sqrt{e}}(\sqrt{e}\mathbf{L}^k)_{,k} = \bar{\epsilon}^{ijk}(\hat{T}^n_{ji,k} + \hat{T}^l_{ji}\hat{\Gamma}^n_{lk})\mathbf{m}_n.$$

This is equivalent to

$$\hat{\epsilon}^{ijk}(\hat{T}^n_{ji,k} + \hat{T}^l_{ji}\hat{\Gamma}^n_{lk}) = 0. \tag{$\star$}$$

(c) Next, note that

$$\begin{aligned}
\hat{\epsilon}^{mlj}\hat{R}^k_{\cdot mlj} &= \hat{\epsilon}^{mlj}\left[\hat{\Gamma}^k_{mj,l} - \hat{\Gamma}^k_{ml,j} - \left(\hat{\Gamma}^k_{ij}\hat{\Gamma}^i_{ml} - \hat{\Gamma}^k_{il}\hat{\Gamma}^i_{mj}\right)\right] \\
&= \hat{\epsilon}^{mlj}\left[\hat{\Gamma}^k_{[mj],l} - \hat{\Gamma}^k_{[ml],j} - \left(\hat{\Gamma}^k_{ij}\hat{\Gamma}^i_{[ml]} - \hat{\Gamma}^k_{il}\hat{\Gamma}^i_{[mj]}\right)\right] \\
&= \hat{\epsilon}^{mlj}\left[\hat{T}^k_{mj,l} - \hat{T}^k_{ml,j} - \left(\hat{\Gamma}^k_{ij}\hat{T}^i_{ml} - \hat{\Gamma}^k_{il}\hat{T}^i_{mj}\right)\right] \\
&= \hat{\epsilon}^{jml}\left(\hat{T}^k_{mj,l} + \hat{T}^i_{mj}\hat{\Gamma}^k_{il}\right) + \hat{\epsilon}^{lmj}\left(\hat{T}^k_{ml,j} + \hat{T}^i_{ml}\hat{\Gamma}^k_{ij}\right) \\
&= 2\hat{\epsilon}^{jml}\left(\hat{T}^k_{mj,l} + \hat{T}^i_{mj}\hat{\Gamma}^k_{il}\right),
\end{aligned}$$

and therefore ($\star$) yields $\hat{\epsilon}^{mlj}\hat{R}^k_{\cdot mlj} = 0$, which is equivalent to

$$\hat{\epsilon}^{mlj}\hat{R}_{kmlj} = 0.$$

Conversely, we prove ($\star$) by reversing the steps and invoking $\hat{R}^k_{\cdot m(lj)} = 0$.

(d) From (5.57) we have

$$\hat{R}_{kmlj} = \hat{\epsilon}_{pkm}\hat{\epsilon}_{qlj}\hat{\Pi}^{pq}.$$

Then,

$$\begin{aligned}
0 &= \hat{\epsilon}^{mlj}\hat{R}_{kmlj} = \hat{\epsilon}^{mlj}\hat{\epsilon}_{mpk}\hat{\epsilon}_{qlj}\hat{\Pi}^{pq} \\
&= \left(\delta^l_p\delta^j_k - \delta^j_p\delta^l_k\right)\hat{\epsilon}_{qlj}\hat{\Pi}^{pq} \\
&= \left(\hat{\epsilon}_{qpk} - \hat{\epsilon}_{qkp}\right)\hat{\Pi}^{pq} \\
&= 2\hat{\epsilon}_{qpk}\hat{\Pi}^{pq}.
\end{aligned}$$

Accordingly, the equations $\hat{\Pi}^{[ij]} = 0$ are identities.

6.6. Replace **G** by **F** and $\mathbf{m}_j$ by $\mathbf{g}_j = \mathbf{y}_{,j}$ in (5.43).

6.9. From $\mathcal{J}_H\mathbf{T} = \mathbf{H}\hat{\mathbf{S}}\mathbf{H}^t$ and $\hat{\mathbf{S}} = \mathcal{C}[\hat{\mathbf{E}}] + o(|\hat{\mathbf{E}}|)$ we have $\mathcal{J}_H tr\mathbf{T} = tr(\mathbf{H}\hat{\mathbf{S}}\mathbf{H}^t) = tr(\mathbf{H}^t\mathbf{H}\hat{\mathbf{S}}) = tr[(\mathbf{I} + 2\hat{\mathbf{E}})\hat{\mathbf{S}}] = tr\hat{\mathbf{S}} + o(|\hat{\mathbf{E}}|)$. Use $\mathcal{J}_H^2 = \det(\mathbf{I} + 2\hat{\mathbf{E}}) = 1 + O(|\hat{\mathbf{E}}|)$ to get $\mathcal{J}_H = 1 + O(|\hat{\mathbf{E}}|)$. Then, $\mathcal{J}_H^{-1} = 1 + O(|\hat{\mathbf{E}}|)$ also, and $tr\mathbf{T} = tr\hat{\mathbf{S}} + o(|\hat{\mathbf{E}}|)$.

6.10. First, note that

$$
\begin{aligned}
E_{11}E_{22} + E_{11}E_{33} + E_{22}E_{33} &= \bar{E}_{11}\bar{E}_{22} + \bar{E}_{11}\bar{E}_{33} + \bar{E}_{22}\bar{E}_{33} + \tfrac{1}{3}\left(tr\hat{\mathbf{E}}\right)^2 \\
&\quad + \tfrac{2}{3}\left(tr\hat{\mathbf{E}}\right)\left(\bar{E}_{11} + \bar{E}_{22} + \bar{E}_{33}\right) \\
&= \tfrac{1}{3}\left(tr\hat{\mathbf{E}}\right)^2 + \tfrac{1}{2}\left(2\bar{E}_{11}\bar{E}_{22} + 2\bar{E}_{11}\bar{E}_{33} + 2\bar{E}_{22}\bar{E}_{33}\right) \\
&= \tfrac{1}{3}\left(tr\hat{\mathbf{E}}\right)^2 + \tfrac{1}{2}\left[\bar{E}_{11}\left(\bar{E}_{22} + \bar{E}_{33}\right) + \bar{E}_{22}\left(\bar{E}_{11} + \bar{E}_{33}\right)\right. \\
&\quad \left. + \bar{E}_{33}\left(\bar{E}_{11} + \bar{E}_{22}\right)\right] \\
&= \tfrac{1}{3}\left(tr\hat{\mathbf{E}}\right)^2 - \tfrac{1}{2}\left(\bar{E}_{11}^2 + \bar{E}_{22}^2 + \bar{E}_{33}^2\right).
\end{aligned}
$$

Thus, the strain energy is expressible in the form

$$U(\hat{\mathbf{E}}) = \tfrac{1}{2}[C_1(E_{11} + E_{22} + E_{33})^2 + C_2(\bar{E}_{11}^2 + \bar{E}_{22}^2 + \bar{E}_{33}^2) + C_3(E_{12}^2 + E_{13}^2 + E_{23}^2)].$$

If this is to be positive-definite, i.e., $U(\hat{\mathbf{E}}) > 0$ for *all* non-zero $\hat{\mathbf{E}}$, it must be positive for non-zero strains with $Dev\hat{\mathbf{E}} = \mathbf{0}$. Then $\frac{1}{2}C_1(tr\hat{\mathbf{E}})^2 > 0$ with $tr\hat{\mathbf{E}} \neq 0$, implying that $C_1 > 0$. Similarly, we can choose $\hat{\mathbf{E}} = Dev\hat{\mathbf{E}}$, with $\bar{E}_{ij} = 0$ for $i \neq j$, to conclude that $C_2 > 0$; and finally, choose $\hat{\mathbf{E}} = Dev\hat{\mathbf{E}}$ with $\bar{E}_{ij} = 0$ for $i = j$, to conclude that $C_3 > 0$. Thus, $C_{1,2,3} > 0$ are necessary conditions for the positive definiteness of $U(\hat{\mathbf{E}})$. From the above expression for $U(\hat{\mathbf{E}})$, they are clearly also sufficient.

To obtain the stress, we use $\hat{\mathbf{S}} \cdot (\hat{\mathbf{E}})^\cdot = \dot{U}$, where

$$
\begin{aligned}
\dot{U} &= C_1(tr\hat{\mathbf{E}})\mathbf{I} \cdot (\hat{\mathbf{E}})^\cdot + C_2[\bar{E}_{11}(\bar{E}_{11})^\cdot + \bar{E}_{22}(\bar{E}_{22})^\cdot + \bar{E}_{33}(\bar{E}_{33})^\cdot] + C_3[E_{12}(E_{12})^\cdot \\
&\quad + E_{13}(E_{13})^\cdot + E_{23}(E_{23})^\cdot].
\end{aligned}
$$

Use $\dot{E}_{ij} = \mathbf{l}_i \cdot (\hat{\mathbf{E}})^\cdot \mathbf{l}_j = \mathbf{l}_i \otimes \mathbf{l}_j \cdot (\hat{\mathbf{E}})^\cdot = Sym(\mathbf{l}_i \otimes \mathbf{l}_j) \cdot (\hat{\mathbf{E}})^\cdot$, and, similarly,

$$
\begin{aligned}
(\bar{E}_{ij})^\cdot &= Sym(\mathbf{l}_i \otimes \mathbf{l}_j) \cdot (Dev\hat{\mathbf{E}})^\cdot \\
&= Sym(\mathbf{l}_i \otimes \mathbf{l}_j) \cdot Dev(\hat{\mathbf{E}})^\cdot = Dev[Sym(\mathbf{l}_i \otimes \mathbf{l}_j)] \cdot Dev(\hat{\mathbf{E}})^\cdot \\
&= Dev[Sym(\mathbf{l}_i \otimes \mathbf{l}_j)] \cdot (\hat{\mathbf{E}})^\cdot,
\end{aligned}
$$

to conclude that

$$(\bar{E}_{11})^\cdot = Dev(\mathbf{l}_1 \otimes \mathbf{l}_1) \cdot (\hat{\mathbf{E}})^\cdot = (\mathbf{l}_1 \otimes \mathbf{l}_1 - \tfrac{1}{3}\mathbf{I}) \cdot (\hat{\mathbf{E}})^\cdot,$$

and, similarly, that

$$(\bar{E}_{22})^{\cdot} = (\mathbf{l}_2 \otimes \mathbf{l}_2 - \tfrac{1}{3}\mathbf{I}) \cdot (\hat{\mathbf{E}})^{\cdot} \quad \text{and} \quad (\bar{E}_{33})^{\cdot} = (\mathbf{l}_3 \otimes \mathbf{l}_3 - \tfrac{1}{3}\mathbf{I}) \cdot (\hat{\mathbf{E}})^{\cdot}.$$

Then,

$$\begin{aligned} \bar{E}_{11}(\bar{E}_{11})^{\cdot} + \bar{E}_{22}(\bar{E}_{22})^{\cdot} + \bar{E}_{33}(\bar{E}_{33})^{\cdot} &= (\bar{E}_{11}\mathbf{l}_1 \otimes \mathbf{l}_1 + \bar{E}_{22}\mathbf{l}_2 \otimes \mathbf{l}_2 + \bar{E}_{33}\mathbf{l}_3 \otimes \mathbf{l}_3) \cdot (\hat{\mathbf{E}})^{\cdot} \\ &\quad -\tfrac{1}{3}(\bar{E}_{11} + \bar{E}_{22} + \bar{E}_{33})\mathbf{I} \cdot (\hat{\mathbf{E}})^{\cdot} \\ &= (\bar{E}_{11}\mathbf{l}_1 \otimes \mathbf{l}_1 + \bar{E}_{22}\mathbf{l}_2 \otimes \mathbf{l}_2 + \bar{E}_{33}\mathbf{l}_3 \otimes \mathbf{l}_3) \cdot (\hat{\mathbf{E}})^{\cdot}. \end{aligned}$$

Thus,

$$\begin{aligned} \dot{U} &= C_1(tr\hat{\mathbf{E}})\mathbf{I} \cdot (\hat{\mathbf{E}})^{\cdot} + C_2(\bar{E}_{11}\mathbf{l}_1 \otimes \mathbf{l}_1 + \bar{E}_{22}\mathbf{l}_2 \otimes \mathbf{l}_2 + \bar{E}_{33}\mathbf{l}_3 \otimes \mathbf{l}_3) \cdot (\hat{\mathbf{E}})^{\cdot} \\ &\quad + C_3[E_{12}Sym(\mathbf{l}_1 \otimes \mathbf{l}_2) + E_{13}Sym(\mathbf{l}_1 \otimes \mathbf{l}_3) + E_{23}Sym(\mathbf{l}_2 \otimes \mathbf{l}_3)] \cdot (\hat{\mathbf{E}})^{\cdot}, \end{aligned}$$

and therefore

$$\begin{aligned} \hat{\mathbf{S}} &= C_1(tr\hat{\mathbf{E}})\mathbf{I} + C_2(\bar{E}_{11}\mathbf{l}_1 \otimes \mathbf{l}_1 + \bar{E}_{22}\mathbf{l}_2 \otimes \mathbf{l}_2 + \bar{E}_{33}\mathbf{l}_3 \otimes \mathbf{l}_3) \\ &\quad + \tfrac{1}{2}C_3[E_{12}(\mathbf{l}_1 \otimes \mathbf{l}_2 + \mathbf{l}_2 \otimes \mathbf{l}_1) + E_{13}(\mathbf{l}_1 \otimes \mathbf{l}_3 + \mathbf{l}_3 \otimes \mathbf{l}_1) + E_{23}(\mathbf{l}_2 \otimes \mathbf{l}_3 + \mathbf{l}_3 \otimes \mathbf{l}_2)]. \end{aligned}$$

7.3. From Problem 6.10 we have

$$\tilde{F}(Dev\hat{\mathbf{S}}) = \tfrac{1}{2}A_1(\bar{S}_{11}^2 + \bar{S}_{22}^2 + \bar{S}_{33}^2) + A_2(S_{12}^2 + S_{13}^2 + S_{23}^2) + const.,$$

where $A_{1,2}$ are constants and the stress components are with respect to the normalized lattice basis $\{\mathbf{l}_i\}$. As in that problem, we also have

$$\begin{aligned} F_{\hat{\mathbf{S}}} &= A_1(\bar{S}_{11}\mathbf{l}_1 \otimes \mathbf{l}_1 + \bar{S}_{22}\mathbf{l}_2 \otimes \mathbf{l}_2 + \bar{S}_{33}\mathbf{l}_3 \otimes \mathbf{l}_3) \\ &\quad + A_2[S_{12}(\mathbf{l}_1 \otimes \mathbf{l}_2 + \mathbf{l}_2 \otimes \mathbf{l}_1) + S_{13}(\mathbf{l}_1 \otimes \mathbf{l}_3 + \mathbf{l}_3 \otimes \mathbf{l}_1) + S_{23}(\mathbf{l}_2 \otimes \mathbf{l}_3 + \mathbf{l}_3 \otimes \mathbf{l}_2)]. \end{aligned}$$

Note that $F_{\hat{\mathbf{S}}} \in Dev$. Then,

$$\hat{\mathbf{S}} \cdot F_{\hat{\mathbf{S}}} = Dev\hat{\mathbf{S}} \cdot F_{\hat{\mathbf{S}}} = A_1(\bar{S}_{11}^2 + \bar{S}_{22}^2 + \bar{S}_{33}^2) + 2A_2(S_{12}^2 + S_{13}^2 + S_{23}^2),$$

where we've used $\bar{S}_{ij} = S_{ij}$ for $i \neq j$. Using

$$\left|Dev\hat{\mathbf{S}}\right|^2 = \bar{S}_{11}^2 + \bar{S}_{22}^2 + \bar{S}_{33}^2 + 2(S_{12}^2 + S_{13}^2 + S_{23}^2),$$

we see that $\hat{\mathbf{S}} \cdot F_{\hat{\mathbf{S}}}$ vanishes if $Dev\hat{\mathbf{S}}$ vanishes (because, then, $\bar{S}_{11}^2 + \bar{S}_{22}^2 + \bar{S}_{33}^2 = 0$ and $S_{12}^2 + S_{13}^2 + S_{23}^2 = 0$). Further, to ensure that $\hat{\mathbf{S}} \cdot F_{\hat{\mathbf{S}}} > 0$ when $Dev\hat{\mathbf{S}} \neq 0$ it is necessary and sufficient that $A_1 > 0$ (pick $S_{ij} = 0$ for $i \neq j$) and $A_2 > 0$ (pick $\bar{S}_{11} = \bar{S}_{22} = 0$; thus, $\bar{S}_{33} = 0$ too).

Because $A_2 > 0$, we can rewrite the yield condition $\tilde{F}(Dev\hat{\mathbf{S}}) = 0$ with $\tilde{F}$ given by

$$\tilde{F}(Dev\hat{\mathbf{S}}) = \tfrac{1}{2}A(\bar{S}_{11}^2 + \bar{S}_{22}^2 + \bar{S}_{33}^2) + S_{12}^2 + S_{13}^2 + S_{23}^2 - k^2,$$

where A is a positive constant and k is the yield stress for stress states of the form $S_{12}(\mathbf{l}_1 \otimes \mathbf{l}_2 + \mathbf{l}_2 \otimes \mathbf{l}_1)$, $S_{13}(\mathbf{l}_1 \otimes \mathbf{l}_3 + \mathbf{l}_3 \otimes \mathbf{l}_1)$ or $S_{23}(\mathbf{l}_2 \otimes \mathbf{l}_3 + \mathbf{l}_3 \otimes \mathbf{l}_2)$, i.e., for shearing stresses on the crystallographic axes. As expected, the symmetry of the lattice implies that the yield stresses on these axes are all equal.

Note that this yield function reduces to von Mises' function when $A = 1$.

7.4. We have

$$\dot{\mathbf{G}}\mathbf{G}^{-1} = \lambda F_{\hat{\mathbf{S}}},$$

where

$$F(\hat{\mathbf{S}}) = \tfrac{1}{2}Dev\hat{\mathbf{S}} \cdot Dev\hat{\mathbf{S}} - [\alpha(tr\hat{\mathbf{S}}) + k]^2.$$

Then,

$$\begin{aligned} F_{\hat{\mathbf{S}}} \cdot (\hat{\mathbf{S}})^{\cdot} &= Dev\hat{\mathbf{S}} \cdot Dev(\hat{\mathbf{S}})^{\cdot} - 2\alpha[\alpha(tr\hat{\mathbf{S}}) + k]\mathbf{I} \cdot (\hat{\mathbf{S}})^{\cdot} \\ &= Dev\hat{\mathbf{S}} \cdot (\hat{\mathbf{S}})^{\cdot} - 2\alpha[\alpha(tr\hat{\mathbf{S}}) + k]\mathbf{I} \cdot (\hat{\mathbf{S}})^{\cdot}, \end{aligned}$$

yielding

$$F_{\hat{\mathbf{S}}} = Dev\hat{\mathbf{S}} - 2\alpha[\alpha(tr\hat{\mathbf{S}}) + k]\mathbf{I}.$$

In the rigid-plastic case we have $\mathbf{H} = \mathbf{Q} \in Orth^+$, $\hat{\mathbf{S}} = \mathbf{Q}^t\mathbf{T}\mathbf{Q}$, $Dev\hat{\mathbf{S}} = \mathbf{Q}^t\boldsymbol{\tau}\mathbf{Q}$, where $\boldsymbol{\tau} = dev\mathbf{T}$. Then $tr\hat{\mathbf{S}} = tr\mathbf{T}$, and

$$\begin{aligned} \mathbf{D} &= \mathbf{Q}(\dot{\mathbf{G}}\mathbf{G}^{-1})\mathbf{Q}^t = \lambda\mathbf{Q}F_{\hat{\mathbf{S}}}\mathbf{Q}^t \\ &= \lambda\{\boldsymbol{\tau} - 2\alpha[\alpha(tr\mathbf{T}) + k]\mathbf{i}\}, \end{aligned}$$

which implies that

$$tr\mathbf{D} = -6\alpha\lambda[\alpha(tr\mathbf{T}) + k].$$

This is generally non-zero, and so the material is compressible.

7.6. We have $\mathbf{v} = w(r)\mathbf{k}$ and $\mathbf{L} = \mathbf{k} \otimes gradw = w'(r)\mathbf{k} \otimes \mathbf{e}_r$. Then, $\dot{\mathbf{v}} = \mathbf{v}_t + \mathbf{L}\mathbf{v} = \mathbf{0}$, and we solve $div\mathbf{T} = \mathbf{0}$, assuming zero body force, i.e.,

$$gradp = div\boldsymbol{\tau},$$

with

$$\boldsymbol{\tau} = (2\eta + \tfrac{\sqrt{2}k}{|\mathbf{D}|})\mathbf{D}, \quad \text{if} \quad |\boldsymbol{\tau}| \geq 2k,$$

and $\mathbf{D} = \mathbf{0}$ if $|\boldsymbol{\tau}| < 2k$, where

$$\mathbf{D} = \tfrac{1}{2}w'(r)(\mathbf{k} \otimes \mathbf{e}_r + \mathbf{e}_r \otimes \mathbf{k}).$$

Thus, $|\mathbf{D}| = \frac{1}{2}(w')^2$ and

$$\boldsymbol{\tau} = \tau(\mathbf{k} \otimes \mathbf{e}_r + \mathbf{e}_r \otimes \mathbf{k}),$$

where

$$\tau(r) = \eta w' + kw'/\left|w'\right|.$$

We thus have

$$grad p = (\tau' + \tau/r)\mathbf{k},$$

implying that $\partial p/\partial r = 0 = \partial p/\partial \varphi$ and

$$p'(z) = \tau' + \tau/r.$$

Thus, $p''(z) = 0$ and $p'(z)$, the axial pressure gradient, is constant. If this is zero, then

$$\tau(r) = A/r,$$

where A is constant. We have $|\boldsymbol{\tau}| = \sqrt{2}\tau$ and so yield occurs when $|\tau(r)| \geq k$, i.e., for values of radius satisfying

$$r \leq |A|/k.$$

From the relation between τ and w we have

$$\eta w' + kw'/\left|w'\right| = \tau = A/r.$$

This is valid for $r \leq |A|/k$. Given the geometry of the problem, when $W > 0$ it is reasonable to expect that $w' < 0$. Then,

$$\eta w' - k = A/r,$$

which gives

$$\eta w(r) = kr + A \ln r + B,$$

where B is another constant. This is valid in the region $(a, \bar{r})$, where

$$\bar{r} = |A|/k.$$

In the region $(\bar{r}, b)$ we have $|\tau(r)| < k$ and $|\mathbf{D}| = 0$; thus, $w' = 0$ in this region, implying that $w(\bar{r}) = w(b) = 0$. Using this together with $w(a) = W(> 0)$, we obtain

$$B = -k\bar{r} - A\ln\bar{r} \quad \text{and} \quad A\ln(\bar{r}/a) = -\eta W - k(\bar{r} - a).$$

The latter gives

$$A/k = [-(\eta/k)W - (\bar{r} - a)]/\ln(\bar{r}/a),$$

which implies that $A < 0$. Using $\bar{r} = -A/k$, we can reduce this to

$$x\ln x = (\eta/ka)W + x - 1,$$

where $x = \bar{r}/a$. This determines x (hence $\bar{r}$) in terms of W, or W in terms of x. Writing $W(x)$ in the latter case, we find that

$$(\eta/ka)W'(x) = \ln x > 0.$$

Thus, x increases with increasing W, as expected, until $W = W^*$, corresponding to $x = b/a$. For $W \geq W^*$ (corresponding to yielding of the entire annulus) we put $\bar{r} = b$ in the above expressions to determine A and B in terms of W.

7.17. We have

$$\begin{aligned} T_{rr,r} + \tfrac{1}{r}(T_{rr} - T_{\varphi\varphi} + T_{r\varphi,\varphi}) &= 0, \\ T_{r\varphi,r} + \tfrac{1}{r}(2T_{r\varphi} + T_{\varphi\varphi,\varphi}) &= 0 \end{aligned}$$

and

$$(T_{rr} - T_{\varphi\varphi})^2 + 4T_{r\varphi}^2 = 4k^2.$$

With a little effort you can rewrite the first pair of equations as

$$\begin{aligned} (rT_{rr})_{,r} + T_{r\varphi,\varphi} &= T_{\varphi\varphi}, \\ T_{\varphi\varphi,\varphi} + \tfrac{1}{r}(r^2T_{r\varphi})_{,r} &= 0. \end{aligned} \qquad (\star)$$

Substitute the first of these into the second to get

$$(rT_{rr})_{,r\varphi} + T_{r\varphi} + (rT_{r\varphi})_{,r} + T_{r\varphi,\varphi\varphi} = 0.$$

If $T_{r\varphi,r} = 0$ this simplifies to

$$(rT_{rr})_{,r\varphi} + 2T_{r\varphi} + T_{r\varphi,\varphi\varphi} = 0.$$

Use the yield condition to write

$$rT_{rr} = r(T_{rr} - T_{\varphi\varphi}) + rT_{\varphi\varphi} = \pm 2r\sqrt{k^2 - T_{r\varphi}^2} + rT_{\varphi\varphi}.$$

We choose the positive root for definiteness. The solution for the negative root proceeds similarly.

With $T_{r\varphi,r} = 0$ we get

$$(rT_{rr})_{,r\varphi} = 2(\sqrt{k^2 - T_{r\varphi}^2})_{,\varphi} + (rT_{\varphi\varphi,\varphi})_{,r},$$

where, from (⋆), $rT_{\varphi\varphi,\varphi} = -(r^2 T_{r\varphi})_{,r} = -2rT_{r\varphi}$. Then, $(rT_{\varphi\varphi,\varphi})_{,r} = -2T_{r\varphi}$ and the first of the last three equations becomes

$$T_{r\varphi,\varphi\varphi} = -2(\sqrt{k^2 - T_{r\varphi}^2})_{,\varphi}.$$

Integrate once to get

$$T_{r\varphi,\varphi} = -2\sqrt{k^2 - T_{r\varphi}^2} + const.$$

One family of solutions can be obtained by taking the constant to be zero. The resulting equation has the solution

$$T_{r\varphi} = -k\sin(c + 2\varphi),$$

where c is another constant. Using $T_{\varphi\varphi,\varphi} = -2T_{r\varphi}$ (from (⋆)), we can integrate again to get $T_{\varphi\varphi}$, and then use the yield condition to get T_{rr}.

Note that another class of solutions, consistent with the assumption $T_{r\varphi,r} = 0$, is $T_{r\varphi} = k$ (or $-k$). In this case the slip lines are aligned with the radial and azimuthal axes. The yield condition then requires $T_{\varphi\varphi} = T_{rr}$, and (⋆) gives

$$T_{\varphi\varphi} = T_{rr} = -2k\varphi + const.$$

This solution plays an important role in understanding the stress field near a corner of a rectangular block. See the interesting discussion on pp. 541 and 542 of Nadai's book *Theory of Flow and Fracture of Solids*.

7.19. The boundary conditions give

$$\tau\mathbf{k} = \mathbf{T}\mathbf{n} = p\mathbf{e}_r - k(\mathbf{t}\cdot\mathbf{e}_r)\mathbf{k},$$

so that

$$\tau/k = -\mathbf{e}_r \cdot \mathbf{t}_{|r=a}.$$

This makes sense because $|\tau/k| \leq 1$ and $\mathbf{t}$ is a unit vector.

To get an axisymmetric stress field, assume that

$$\mathbf{t} = t_r(r)\mathbf{e}_r + t_\varphi(r)\mathbf{e}_\varphi,$$

with

$$t_r^2 + t_\varphi^2 = 1.$$

With a little work you can show that

$$grad\mathbf{t} = t_r'\mathbf{e}_r \otimes \mathbf{e}_r + \tfrac{1}{r}t_r\mathbf{e}_\varphi \otimes \mathbf{e}_\varphi + t_\varphi'\mathbf{e}_\varphi \otimes \mathbf{e}_r - \tfrac{1}{r}t_\varphi\mathbf{e}_r \otimes \mathbf{e}_\varphi,$$

and

$$div\mathbf{t} = tr(grad\mathbf{t}) = t_r' + \tfrac{1}{r}t_r.$$

Because $div\mathbf{t} = 0$ and in view of the boundary condition, we then have

$$t_r = -(a/r)\tau/k$$

and

$$t_\varphi = \pm\sqrt{1-(a/r)^2(\tau/k)^2}$$

for $r \geq a$. Evidently $t_r \to 0$ and $t_\varphi \to \pm 1$ as $r/a \to \infty$, so that $\mathbf{t} \to \pm\mathbf{e}_\varphi$ far away from the hole boundary.

To characterize a velocity field consistent with this solution, recall that

$$\mathbf{s} \cdot grad w = 0,$$

where

$$\mathbf{s} = \mathbf{k} \times \mathbf{t} = t_r\mathbf{e}_\varphi - t_\varphi\mathbf{e}_r \quad \text{and} \quad grad w = (\partial w/\partial r)\mathbf{e}_r + \tfrac{1}{r}(\partial w/\partial\varphi)\mathbf{e}_\varphi.$$

Then $w(r, \varphi)$ satisfies the linear differential equation

$$t_r'\partial w/\partial\varphi + t_\varphi\partial w/\partial r = 0.$$

To be admissible, any solution to this equation must be such that $\lambda \geq 0$, where

$$\lambda\hat{\tau} = \tfrac{1}{2}grad w.$$

With $\hat{\tau} = k\mathbf{t}$, this requires that $\mathbf{t} \cdot grad w \geq 0$, i.e.,

$$rt_r\partial w/\partial r + t_\varphi\partial w/\partial\varphi \geq 0.$$

8.1. This follows immediately from the invariance of $D\mathbf{u} = (u^j\mathbf{e}_j)_{,i} \otimes \mathbf{e}^i = u^j_{|i}\mathbf{e}_j \otimes \mathbf{e}^i$. We also have $\bar{\Gamma}^{*k}_{ij} = \bar{\Gamma}^k_{ij}$ because both connections are Levi-Civita and the starred and unstarred metrics coincide.

8.12. Differentiate the result of Problem 8.1.

9.5. We have $\dot{\mathbf{M}} = \dot{\mathbf{G}}^t\mathbf{G} + \mathbf{G}^t\dot{\mathbf{G}}$. Use $\dot{\mathbf{G}} = (\dot{\mathbf{G}}\mathbf{G}^{-1})\mathbf{G}$ to write this as $\dot{\mathbf{M}} = 2\mathbf{G}^t[Sym(\dot{\mathbf{G}}\mathbf{G}^{-1})]\mathbf{G}$. If $\dot{\mathbf{G}}\mathbf{G}^{-1} \in Sym$, then $Sym(\dot{\mathbf{G}}\mathbf{G}^{-1}) = \dot{\mathbf{G}}\mathbf{G}^{-1}$ and the inverse relation is $2\dot{\mathbf{G}} = \mathbf{G}^{-t}\dot{\mathbf{M}}$. The flow rule (7.8) for isotropic materials is thus easily seen to be equivalent to $\dot{\mathbf{M}} = 2\lambda\mathbf{G}^t(F_{\hat{\mathbf{S}}})\mathbf{G}$.

For von Mises' yield function, Eq. (7.13) may be combined with $\mathcal{J}_G\hat{\mathbf{S}} = \mathbf{G}\mathbf{S}\mathbf{G}^t$, where $\mathbf{S}$ is the Piola–Kirchhoff stress relative to κ, to reduce this to $\dot{\mathbf{M}}\mathbf{M}^{-1} = \mu Dev(\mathbf{M}\mathbf{S})$, where $\mu = 2\lambda/\sqrt{\mathcal{J}_M}$. Using $\dot{\mathcal{J}}_M = \mathbf{M}^* \cdot \dot{\mathbf{M}} = \mathcal{J}_M tr(\dot{\mathbf{M}}\mathbf{M}^{-1})$, we conclude that $\dot{\mathcal{J}}_M = 0$ and hence that this form of the flow rule preserves plastic incompressibility.

9.6. No, the $\{^{\,k}_{ij}\}$ are not tensorial and so the object $\{^{\,k}_{ij}\}\mathbf{m}_k \otimes \mathbf{m}^i \otimes \mathbf{m}^j$ can be made to assume arbitrary values simply by changing the coordinate system.

9.7. To the author's knowledge this is an open problem.

9.8. With reference to Problem 9.5, allow the metric m_{ij}, and hence $\mathbf{G}$, to vary with time while holding its partial coordinate derivatives fixed at the point in question.

9.10. With $w^j = \boldsymbol{\mu}^{ij} \cdot \dot{\mathbf{m}}_i$ we have

$$\begin{aligned}
\boldsymbol{\mu}^{ij} \cdot \dot{\mathbf{m}}_{i|j} &= \boldsymbol{\mu}^{ij} \cdot (\dot{\mathbf{m}}_{i,j} - \dot{\mathbf{m}}_k\bar{\Gamma}^k_{ij}) = w^j_{,j} - \boldsymbol{\mu}^{ij}_{,j} \cdot \dot{\mathbf{m}}_i - \boldsymbol{\mu}^{ij} \cdot \dot{\mathbf{m}}_k\bar{\Gamma}^k_{ij} \\
&= w^j_{|j} - w^k\bar{\Gamma}^j_{kj} - \boldsymbol{\mu}^{ij}_{,j} \cdot \dot{\mathbf{m}}_i - \boldsymbol{\mu}^{ij} \cdot \dot{\mathbf{m}}_k\bar{\Gamma}^k_{ij} \\
&= (\boldsymbol{\mu}^{ij} \cdot \dot{\mathbf{m}}_i)_{|j} - (\boldsymbol{\mu}^{ij}_{,j} + \boldsymbol{\mu}^{ik}\bar{\Gamma}^j_{kj} + \boldsymbol{\mu}^{kj}\bar{\Gamma}^i_{kj}) \cdot \dot{\mathbf{m}}_i.
\end{aligned}$$

9.13. We have $\mathbf{N} = \mathbf{G}^t W_{\boldsymbol{\alpha}}$. Because $\boldsymbol{\alpha} \to \mathbf{R}^t\boldsymbol{\alpha}\mathbf{R}$, it follows, exactly as in the derivation of (6.28), that $W_{\boldsymbol{\alpha}} \to \mathbf{R}^t W_{\boldsymbol{\alpha}}\mathbf{R}$. From $\mathbf{G} \to \mathbf{R}^t\mathbf{G}$ we then have that $\mathbf{N} \to \mathbf{N}\mathbf{R}$, and the definition of $\boldsymbol{\eta}^{ij}$ in (9.84) yields $\boldsymbol{\eta}^{ij} \to \mathbf{R}^t\boldsymbol{\eta}^{ij}$. From (5.43) we have $Curl\dot{\mathbf{G}} = \bar{\varepsilon}^{ijk}\mathbf{e}_k \otimes \dot{\mathbf{m}}_{[j,i]}$. Then $\dot{\mathbf{G}} \to \mathbf{R}^t\dot{\mathbf{G}}$ yields $\dot{\mathbf{m}}_j \to \mathbf{R}^t\dot{\mathbf{m}}_j$, $\mathbf{e}_k \to \mathbf{e}_k$, and $Curl\dot{\mathbf{G}} \to \bar{\varepsilon}^{ijk}\mathbf{e}_k \otimes \mathbf{R}^t\dot{\mathbf{m}}_{[j,i]} = (Curl\dot{\mathbf{G}})\mathbf{R}$. Finally, $\mathbf{N} \cdot Curl\dot{\mathbf{G}} \to \mathbf{N}\mathbf{R} \cdot (Curl\dot{\mathbf{G}})\mathbf{R} = \mathbf{N} \cdot Curl\dot{\mathbf{G}}$, this following from the easily derived identity $\mathbf{A}\mathbf{B} \cdot \mathbf{C}\mathbf{D} = \mathbf{A}\mathbf{B}\mathbf{D}^t \cdot \mathbf{C}$ and the orthogonality of $\mathbf{R}$.

Index